ROYAL
OBSERVATORY
GREENWICH

T0075426

Space Exploration

Dhara Patel

Royal Observatory Greenwich
Illuminates

First published in 2021 by the National Maritime Museum, Park Row, Greenwich, London SE10 9NF

ISBN: 978-1-906367-88-6

At the heart of the UNESCO World Heritage Site of Maritime Greenwich are the four world-class attractions of Royal Museums Greenwich – the National Maritime Museum, the Royal Observatory, the Queen's House and *Cutty Sark*.

rmg.co.uk

A CIP catalogue record for this book is available from the British Library.

Typesetting by ePub KNOWHOW
Cover design by Ocky Murray
Diagrams by Dave Saunders
Printed and bound by CPI Group (UK) Ltd, Croydon, CR0 4YY

10 9 8 7 6 5 4 3 2 1

About the Author

Dhara Patel is an astronomer and science communicator at the Royal Observatory Greenwich. Beginning her career as a secondary school science teacher after completing her Masters in Physics, she has since spent the last five years sharing her passion for science and astronomy from the Observatory, as well as at national science festivals and through the press and media.

Entrance of the Royal Observatory, Greenwich, about 1860.

About the Royal Observatory Greenwich

The historic Royal Observatory has stood atop Greenwich Hill since 1675, and documents over 800 years of astronomical observation and timekeeping. It is truly the home of space and time, with the world-famous Greenwich Meridian Line, awe-inspiring astronomy and the Peter Harrison Planetarium. The Royal Observatory is the perfect place to explore the Universe with the help of our very own team of astronomers. Find out more about the site, book a planetarium show, or join one of our workshops or courses online at rmg.co.uk.

Contents

Introduction   1

How Did Space Exploration Begin?   5

The First Satellites in Space   21

Racing to Achieve Human Spaceflight   35

Preparing for Crewed Moon Missions   49

Human Exploration of the Moon   63

Space Stations Through the Decades   82

The Future of Space Missions   108

Glossary   141

Useful Acronyms   145

Introduction

It could be said that the exploration of space through astronomy (or perhaps even through its roots in **astrology**) began with the very earliest humans who looked up at the sky and pondered over what it was that they were seeing. But true space exploration is something that has only developed in the last century or so. Cave paintings, rock carvings, and bones marked with the phases of the Moon date back beyond 10,000 BCE, and the principle of astrology likely originated with the Babylonians in the second millennium BCE. But as the planetary bodies began to be investigated from a scientific and logical standpoint a few thousand years later, by the Greeks in particular, the archaic bond between astrology and astronomy began to snag; astronomy sprouted as its own field. However, others would say that it did not become truly separate until the seventeenth century during the **'Age of Reason'**. Space exploration is yet another branch out of this lineage. It is considered the physical investigation of the Universe beyond the surface of the Earth through the use of spacecraft. A spacecraft is any vehicle (either crewed or uncrewed), or a robotic device designed for travel or operation outside the Earth's atmosphere. From studying our closest planetary neighbours in the Solar System to examining the large active galaxies that electrify their surroundings in the most distant depths of space, spacecraft have helped us learn more about the Universe in the past hundred years than we have in the entirety of human history. And yet there is so much more to unravel.

Space exploration did not suddenly manifest. It grew out of the scientific work of many men and women who theorised the mechanics and motions of celestial bodies using their Earth-bound observations. Space exploration would not be possible if it were not for the empirical knowledge composed by scientific thinkers that lived in bygone centuries. It is from this foundational work that people came to recognise how limited we were by terrestrial observation, and realise the need for exploration through spacecraft to improve our understanding of the vast Universe that we live in.

For a relatively young scientific field, space exploration has a great history and an even greater array of achievements to boast about. Not only have we been able to observe astronomical bodies with greater clarity or even see those that were invisible to us before, but we have also been able to extend the senses we use to probe the Universe. We've used robotic instruments to 'taste' and sent humans into space to touch and return samples from worlds beyond our own.

In the following chapters we'll examine space exploration, beginning with the first spacecrafts and satellites launched above the surface of the Earth, through to current and upcoming missions that push the boundaries of technology, delving deeper into the mysteries of the Universe. But first it's worth looking back to how this field came about. For thousands of years, humans have studied the skies using their eyes. But once advancing scientific understanding illustrated the many limits of human visibility, it became clear that we had to leave the Earth to truly understand the cosmos.

How Did Space Exploration Begin?

For most of humanity's history we have studied space using our eyes. The frustrations of not being able to venture out to the celestial bodies didn't subdue our interest in wanting to find out more about the Universe. In fact, the idea that we were seeing entities beyond the Earth and its atmosphere, made it all the more exciting to investigate them further.

Even with limited tools, over centuries and millennia, scholars and scientists have recorded numerous observations – each new generation building on the work of those before them. Through these detailed reports of what they witnessed when looking into space, some of the mysteries of the Universe, such as the nature and mechanics of our solar system, became apparent to us over time. Ancient civilisations tied their lives to the movement of celestial bodies, using them to underpin religious traditions such as the markings of Easter and Ramadan which were first set in 325 CE and 610 CE respectively, but also Diwali, a festival which stretches back over 2,500 years. On a more practical note, people began to use the position of the Sun to govern the length and time of day, and used the Moon's activity to create calendars and denote months – it became the basis upon which farmers of early societies were guided to sow and harvest their crops. Detailed studies and an understanding of objects in space therefore became incredibly important.

It's remarkable that before the invention of the telescope many of the foundations of astronomy were already laid. As far back as the fourth and fifth centuries BCE, many ancient cultures such as the Egyptians, the Mayans, and later the Babylonians tracked the motions of objects and realised perhaps two of the most fundamental concepts – that stars appeared fixed in the sky and were very far away, and that there were planets which wandered across the sky (some faster and others more glacially), and were likely much closer. The correct understanding of these observations remained a minority view for a very long time. Most people had believed we were living in an Earth-centred Universe, known as the **geocentric** model. But the mathematician and astronomer Nicolaus Copernicus (1473–1543) set the ball rolling for a drastic shift in the accepted theory. Just before his death, he became the first to demonstrate mathematically that the Earth instead was likely to sit within a **heliocentric** model, with the Sun at the centre of our solar system and the planets, including Earth, orbiting around. It took over a century for our current understanding of the Solar System to be widely accepted over the geocentric view held by the Catholic Church, and much of that came about due to the invention of the telescope.

In 1608, a Dutch eyeglass maker called Hans Lippershey (*c.*1570–1619) filed for a patent for his device that he called *kijker*, meaning 'looker'. It was the first record for such an instrument and so he is often credited as the inventor of the telescope, but without doubt there were other spectacle makers experimenting with similar contraptions around the same time. Often Jacob Metius (*c.*1571–*c.*1628) and Zacharias Janssen (*c.*1585–*c.*1632) are names that appear in this dispute. Copies of Lippershey's design were soon shared, and by 1609 the Englishman Thomas Harriot (c.1560–1621) had a 6-times magnification telescope compared to the 3-times magnification telescope designed by Lippershey. By the following year, the father of observational astronomy Galileo Galilei (1564–1642) had designed a telescope with 30-times magnification – a telescope he used to make some revolutionary observations.

Initially unbeknown to him, Galilei observed and discovered four satellites, or moons, around Jupiter which now bear his name. The Galilean moons – Io, Europa, Ganymede and Callisto – were the first objects to be seen orbiting around another planetary body and these observations began to threaten the religious beliefs held by the Roman Catholic Church. It gave plausible evidence for what Copernicus had suggested – that the Earth was not the centre of the Universe, and everything did not revolve around it.

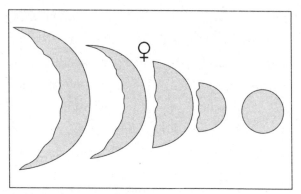

Galileo's drawings of the phases of Venus, published in the seventeenth century. His observations of the planet determined that an Earth-centred system could not be possible.

Galilei also observed Venus and several observations revealed that it appeared to have phases like the Moon – appearing as a small disc when fully illuminated and then apparently growing in size as the phases waned, eventually being seen as a crescent before disappearing from view. These observations could not be explained in an Earth-centred system. Galilei was careful to conceal his revelations (some as anagrams) when sharing with his trusted associates,

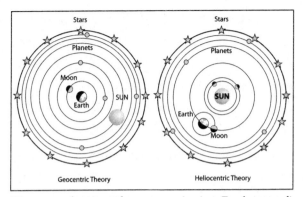

Diagrams showing the geocentric (or Earth-centred) model and the heliocentric (or Sun-centred) model of the structure of the Universe.

but he was eventually tried for heresy and placed under house arrest in 1633 until his death in 1642.

Even though observational evidence had begun to mount, it was with Sir Isaac Newton's (1643–1727) theory of universal gravitation, which he laid out in his *Philosophiae Naturalis Principia Mathematica* (simply referred to as *Principia*) in 1687, that the Sun-centred model proposed by Copernicus was finally cemented. Newton (an English physicist, mathematician, and astronomer) described gravity as the force that kept the planets in orbit around the Sun, eliminating the need for transparent spheres on which the planets were held, as was previously believed.

There were likely to have been early prototypes, however it was Newton that built the first successful reflecting telescope in 1668, which used mirrors instead of lenses to collect and manipulate light. It is an invention that has become the basis for the world's largest telescopes today – but why? Prototype lensed (or refracting) telescopes suffered from chromatic aberration – a problem created from the lenses dispersing the light, preventing some of it from focusing properly. Newton conducted many experiments on light and saw that white light dispersed when passing through a glass prism. He concluded that white light was composed of a **spectrum** of light. The glass prism refracted or changed the direction of light and because different colours are refracted by different amounts, the white light became dispersed into a spectrum. Newton knew that mirrors would not lead to refraction of light and so there should be no chromatic aberration – an experiment to prove this was his motive for building a reflecting telescope.

Telescopes have many key features. One of the most important aspects of a telescope is its aperture – the size of the lens or mirror collecting the light. The bigger you make it, the more light it can collect, and so the fainter (and therefore more distant) an object you can see. Many people began to experiment with building larger telescopes. But larger lenses become heavier and over time the lenses sag, as glass is an **amorphous solid**. Furthermore, lenses can only be held around the rim as the light needs to pass through them to be collected behind. But the majority of the lens' weight lies at the centre as lenses are formed into a **convex** shape, in order to focus the light as necessary. The larger the lenses were made, the more challenging it became to create flawless designs. The largest refracting telescopes today are not much larger than about 1 metre, capable of resolving a foot-long baguette on the International Space Station (ISS) circling 400 kilometres above us, or a 250-metre crater on the Moon orbiting roughly 380,000 kilometres from Earth. By resolving we mean the ability to distinguish the size or shape of an object but not necessarily being able to make out any details. Similar to pixels on a screen, anything smaller than a baguette at a distance of 400 kilometres (as in the example above) would simply be too small (smaller than a pixel) so would not be resolvable. Although the construction of new lensed telescopes of large scale has all but ceased, the surviving large refractors are still used for **astrometry.** They are in fact preferred by some visual observers because their long focal length allows for high magnification, and the closed tube design significantly eliminates air currents within it, avoiding unsteady images.

But because reflecting telescopes use mirrors, they can be fully supported from behind (as light doesn't need to travel through them), and they can be made lighter than their lensed counterparts as it's only the very top layer that needs to be a perfectly smooth, mirrored surface. What it's constructed over doesn't really matter. The advantage of being able to build larger telescopes using mirrors is why many of the world's largest telescopes and those sent into space are reflectors.

From the turn of the twentieth century, telescopes of magnificent size sprouted up (notably across America) and essentially all of those were reflectors. From the 2.5-metre Hooker Telescope in 1917, and the

5.1-metre Hale telescope in 1948, to the more recent two 10-metre Keck telescopes in 1993 and 1996, and the four 8.2-metre telescopes of the European Southern Observatory's (ESO) Very Large Telescope in 1998. The endeavour to go bigger and better continues with the currently under construction 39.3-metre ESO Extremely Large Telescope, due to see first light in 2025.

With advancing scientific understanding of ground-based observing and the development of urban environments, the location for telescopes had to be considered more carefully. Increasing levels of light pollution and the effects of atmospheric conditions drove engineers and scientists to develop observatories in more remote areas – regions that were devoid of light pollution and at higher altitudes to negate most atmospheric effects. Remote locations around the world that had optimum weather conditions were chosen to enable a greater number of clear nights for observing, such as Chile, the Canary Islands and South Africa.

What has perhaps opened up our understanding of space more than anything is the realisation of what light is and how we detect it – all of it. In 1800, the astronomer William Herschel (1738–1822) detected infrared light by placing a thermometer beyond the red end of the spectrum of visible light and noted that here, seemingly out of the light, it had the highest temperature. Inspired by the discovery, the physicist Johann Wilhelm Ritter (1776–1810) went on to ascertain the existence of ultraviolet light – another invisible form of light, but this time beyond the violet end of the visible spectrum. We now understand that visible light (the light we detect with our eyes) is only a very small fraction of the entire family of light known as the **electromagnetic spectrum**, most of which is invisible to our eyes. It consists of the longest wave radio waves, through microwaves, infrared, visible, ultraviolet, X-rays, and eventually the shortest wavelength gamma-rays.

The first extra-terrestrial source of non-visible light to be detected from Earth was radio waves. In 1931 Karl Jansky (1905–50), an American physicist and radio engineer, discovered radio waves emanating from near the centre of our galaxy, the Milky Way. He used a rotating antennae system he built while working for Bell Telephone Laboratories to investigate the origins of static which was affecting their transatlantic radio transmissions. It was almost beyond question that (disregarding visible light) the first light to be discovered from space would be radio waves, as most wavelengths of this light easily penetrate through the Earth's atmosphere making them detectable on the ground. Other forms of light are attenuated somewhat by the Earth's atmosphere, making them difficult to detect on the surface. Some (like gamma-rays, X-rays, a fraction of ultraviolet, and most infrared light) are blocked by the atmosphere almost entirely and are therefore best detected and observed from space. Collecting different types of light from astronomical sources give us a fuller picture of what's going on – it's like being able to use all your senses instead of just one. These different types of light tell us different things about the object they're coming from; each a clue or a piece of evidence that can help unravel the mystery.

American physicist Lyman Spitzer (1914–97) was the first to conceive the idea of telescopes operating in outer space in 1946 and is the namesake of the National Aeronautics and Space Administration's (NASA) Spitzer Space Telescope. Around this time, the first objects were being launched into space at the birth of the Space Race, but it wasn't until two decades later that the first space-based observatory was launched, and Spitzer's concept became a reality.

There is such a wealth of astronomical knowledge that humans have managed to obtain from simply observing here on the Earth. But to observe from space and venture to some of the insanely distant worlds and even beyond, can reveal details about the Universe that we simply wouldn't be able to deduce from here on Earth, or it would take significant improvements in technology and therefore considerable time before we were able to do so. There lies an underlying rationale for the birth of space exploration – to understand as much as we can, as best we can, and as quickly as we can, from a Universe that is outlandishly large and enigmatic.

The First Satellites in Space

Twelve years after the turn of the twentieth century in the small town of Wirsitz, formerly a part of the German Empire but presently found in modern-day Poland, a young man was born who to a great extent set in motion the dawn of space exploration. Wernher von Braun (1912–77) developed a passion for astronomy as a young boy after being gifted a telescope by his mother. He was so fascinated by speed records set by rocketry pioneers that he ended up in police custody aged 12, after attaching fireworks to a toy car which he detonated in a busy public place. While at school he came across Hermann Oberth's

(1894–1989) first draft of *By Rocket into Planetary Space* and even though he'd always been drawn to the idea of space travel, it was from then that he really began to study physics and mathematics in depth, to pursue his interest in becoming a rocket engineer. Von Braun went on to become Oberth's assistant and together they worked on rocketry research.

It was during the Second World War that von Braun caught the attention of the German Army and began working on the development of the V-2 rocket under their control. Smaller rockets had been built in the 1930s but the size and range of the V-2 was a giant leap. It was a single-stage missile, stabilised by fins, and propelled by a fuel of liquid oxygen and ethyl alcohol. The 'V' stood for *Vergeltungswaffe* which meant 'retaliatory weapon' and it was assigned to attack allied cities as vengeance for their bombings on German lands. From September 1944 to March 1945, V-2 rockets were used to target cities, including London and Antwerp, sadly causing a large number of casualties to civilians. However, a few months prior to the attacks, with a vertical test launch on 20 June 1944, a V-2 became the first artificial object to reach space when it crossed the **Kármán line** (which had not been defined at that time). Climbing 176 kilometres above the Earth's surface, it failed to reach orbital velocity so returned to the Earth with an impact. Nevertheless, it had made the first sub-orbital spaceflight.

The Kármán line was an idea originated by the Hungarian-American physicist Theodore von Kármán (1881–1963) who in the 1950s calculated that the air in the atmosphere becomes so thin above a certain altitude, that an aircraft can't generate sufficient lift force to keep it in the air without traveling faster than the orbital speed. Aircraft generate lift to counteract the gravitational force pulling them back to the Earth, by creating a pressure difference above and below their wings, by nature of their design, how they're angled, and the speed of the aircraft. As von Kármán calculated, above 83.6 kilometres the air is too thin to generate a lift force without travelling faster than orbital velocity, at which point the craft could enter Earth orbit. His proposition was that the boundary of space should be defined at the altitude where the forces due to orbital dynamics become more significant than aerodynamic forces, and it was after the war that the Fédération Aéronautique Internationale (FAI) officially set the definition. However, the actual boundary varies, as the calculation is dependent on the atmospheric and spacecraft properties which are not fixed. Thus, the defining altitude could be considered anywhere between roughly 85–100 kilometres. In general, many including NASA consider the Kármán line to be 100 kilometres above the surface of the Earth – this is where space begins.

Unsurprisingly, after the war ended many nations were eager to get hold of the V-2 technology. It was the world's first ballistic missile, with a motor system capable of deriving enough energy to reach space. It also had an automated guidance system which could continuously track the craft's position, and adjust its fins to correct its trajectory if it veered off track. In the post-war era, a rivalry developed between the United States (US) and the Union of Soviet Socialist Republics (USSR), also known as the Soviet Union – a federal socialist state consisting of Russia and a number of surrounding countries. This Cold War, driven on political and economic grounds, also spurred the Americans and the Soviets to retrieve as many of the V-2 rockets as possible, and the engineers and scientists that had worked on them. Although the Soviets claimed the V-2 factory and test range, von Braun surrendered to the Americans when he was captured by them.

Under the United States' National Advisory Committee for Aeronautics (NACA), von Braun assisted in the Bumper program while working for the US Army. This was a research rocket designed to investigate launching techniques and attain record-breaking speeds and altitudes. The Bumper-WAC, as it became known, consisted of the German V-2 rocket and the second stage WAC Corporal which sat on top. After the first minute of flight at high altitude, the V-2 (before shutting down) would provide a bump or boost to the WAC Corporal second stage. It was the first high speed multistage rocket (launching one rocket from another rocket in flight) – a common design used in most present-day launch vehicles. In 1949, the Bumper-WAC became the first rocket to carry a **payload** into space when it rose to an altitude of 393 kilometres. The small payloads carried by the Bumper-WAC rockets included transmitters and receivers that could measure and send information like air temperature, velocity and even cosmic-ray impacts.

The US began its Earth satellite program with Project Orbiter five years later with the aim of placing a scientific satellite into orbit during the International Geophysical Year, which ran from July 1957 until the end of the following year. In this joint venture between the US Army and the US Navy, von Braun and his team were once again enlisted to build on the development of the Bumper-WAC, and design a new launch vehicle that would make orbital flight a reality.

Following a night launch on 5 October 1957, designers, engineers, and technicians listened for signals they so desperately hoped to hear – the beep of a radio transmission from the satellite they'd just propelled into space; the signal from what would become the first ever satellite in orbital flight. Soon after the launch a telegraph was transmitted. 'As result of great, intense work of scientific institutes and design bureaus, the first artificial Earth satellite has been built.' However, this transmission was not sent by the US – it had come from the Telegraph Agency of the Soviet Union. They had pipped the Americans to the post, and successfully launched Sputnik 1 from the Kazakh Soviet Socialist Republic (now known as the Baikonur Cosmodrome).

Back in 1955, just four days after the US officially announced they would launch a satellite during the International Geophysical Year, the Soviets made the same pledge. They scrapped their original complex and ambitious design to work on the development of PS-1 (otherwise known as Sputnik 1), to ensure they could beat the Americans in the race to place a satellite into orbit. Launched atop the R-7 Russian rocket, Sputnik 1 followed an elliptical orbit for three weeks before its batteries depleted and eventually, due to atmospheric drag, it re-entered the Earth's atmosphere, burning up on re-entry in early 1958.

Although the Soviets provided some details of Sputnik 1 (excluding the precise date of launch) prior to it being sent into space, it still came as a great shock when it happened, as very few outside the state took notice of them. During that historic night, the US and other countries were alerted to look out for the satellite as it passed overhead around dawn and dusk, over the coming days. Sputnik 1 was in low Earth orbit (LEO), meaning that it was less than 2,000 kilometres above the surface of the Earth. It was visible to the naked eye at twilight when darkness covered the ground, since the polished surface of the satellite would be high enough to be lit by the Sun, despite its small size (58 cm). Along with professional radio operators, even amateur observers with short wave receivers that could pick up Sputnik 1's radios signals were encouraged to listen out as it passed overhead roughly every 90 minutes, beeping signals as it did. Around the world there was an uneasy reaction – a mixture of fear and amazement. It shattered perceptions of the US as a technological superpower over the USSR. One could not be underwhelmed by the great human progress the Soviets had made.

The United States made haste and revived Project Orbiter, which was then renamed the Explorer Program in response to the threat to national security they perceived from Sputnik 1. It became known as the 'Sputnik crisis'. Sputnik 1 had undoubtedly started the Space Race. Yet before the Americans could launch Explorer 1, the Soviet's launched another satellite, Sputnik 2, just 32 days after Sputnik 1 had carved out its name in history. On board was Laika, a 6-kilogram dog and the first animal to orbit the Earth. Unfortunately, Laika passed away, likely from overheating in the cabin, within a few hours. But on the evening of 31 January 1958, Explorer 1 was launched aboard Juno 1 (a four-stage American booster rocket developed from von Braun's designs), and the spacecraft became the first to detect the inner Van Allen radiation belt during its four-month operational stint, after which it remained in orbit until 1970. The Van Allen radiation belts are a zone of very energetic, charged particles, many of which stem from the Sun's solar wind. They become trapped around the Earth (primarily in two doughnut-shaped belts), due to the influence of our planet's magnetic field. These charged particles are harmful to humans and can damage spacecraft circuitry and sensors.

This important confirmation would bear significance for future spacecraft travelling beyond LEO where they would enter this radiation zone. And beyond that, further hazards from cosmic rays and solar particle events would affect them.

It was after the launch of Explorer 1 that NASA was created by the United States. It was decided that this new federal agency would be in charge of all non-military space activity, and when it began operating on 1 October 1958 it absorbed NACA including its employees, budget, labs, and facilities. The Americans and the Soviets had now placed satellites into orbit, and over the next few years their efforts in developing spacecraft and launch vehicles would continue and magnify. By the end of the decade, the Soviets had launched Luna 1, which surpassed the Earth's escape velocity. It was succeeded by Luna 2, which made the first impact on the Moon, and then by Luna 3, which returned the first ever image of the far side of the Moon in 1959. For almost our entire human history we've only been able to see the near side of the Moon from our Earth-bound position, and it's only for the tiniest fraction of this incomprehensibly long time, in this era of space exploration, that we've been able to admire the Moon in its full glory. Earlier in 1959, NASA had also given us the first photograph of the Earth from orbit with its Explorer 6 satellite, and the following year its Pioneer 5 spacecraft became the first solar space probe, orbiting the Sun between Earth and Venus. Since its last contact a few months after launch, it remains a derelict spacecraft orbiting the Sun to this day.

From the unsavoury beginnings of the V-2 rocket, space exploration began to blossom. Fierce competition between the Americans and Soviets drove technological advancements that only seemed possible in science fiction novels. But this was just the beginning of the Space Race – the Soviets may have won the first battle, but the war was far from over. Simmering beneath all the uncrewed space missions was a burning desire. A desire fuelled by the Cold War, but also from the innate human desire to explore. The testing, developing, and launching of robotic spacecraft in the bygone years were stepping stones to the ultimate achievement – to be the first nation to take a human into space.

Racing to Achieve
Human Spaceflight

Getting a satellite into space is no mean feat, but sending a 'metal can' as opposed to a human is a far less risky prospect, with fewer serious consequences in the event of a failure. In the vacuum of space, a probe doesn't need a containment of oxygen to keep it functioning, and it can deal with greater temperature extremities than a human would otherwise be capable of surviving in. Most importantly, a robotic probe rarely, if ever, needs to return to Earth in working condition – the only precarious stages are getting it into space and then keeping it working while it's up there. Although there would be great disappointment if a satellite ceased to work, the same feeling is in no way comparable to the grief that would be experienced for a person in its place. Human spaceflight is risky at every stage, and great respect should be offered to all the brave people who dare to put their own lives behind a greater cause – the advancement in human achievement and space exploration.

After the launch of the first satellites, NASA established Project Mercury, the first human space programme by the US, with the goal of putting the first person into Earth orbit and returning them safely – preferably before the Soviets. In 1959, the year after the project began, NASA announced the Mercury Seven – the collective name given to the seven men chosen to fly spacecraft for the project. Scott Carpenter (1925–2013), Gordon

Cooper (1927–2004), John Glenn (1921–2016), Gus Grissom (1926–67), Wally Schirra (1923–2007), Alan Shepard (1923–98), and Deke Slayton (1924–93) had created a new profession in the US. A profession that would become the dream of many young children in the decades to follow, and a reality for a select few – an astronaut.

Each Mercury capsule was designed to carry a single crew member and since no one had ever been to space before, selecting the right candidates for the task became an important consideration. It required looking at what the astronauts would be expected to do, and then consider what professions would likely have people skilled in such qualities. Pilots, divers and mountain climbers were all prospective contenders. Eventually the selection was narrowed down to military test pilots as they would have the transferable experience to pilot the spacecraft, and be used to tight security and handling classified information. Seven selection criteria were drawn up and included constraints based on age, height, physical condition, education and piloting experience.

Rivalry between the Soviets and the Americans in the Space Race continued to brew and by January 1959 the USSR had begun planning and preparing for human spaceflight too, but more covertly. The Vostok programme was an equivalent and competitor programme to Project Mercury, and similarly cosmonaut candidates were chosen from aviation backgrounds. Cosmonaut is the name given to Russian space explorers, equivalent to American astronauts. The physicians from the Soviet Air Force argued that pilots would have the relevant skills and experience: intelligence, exposure to high g-forces, and could remain composed in stressful situations. Interestingly, the candidate pilots were not initially told they might be flying a spacecraft.

The Mercury Seven astronauts had already begun training in 1959 immediately after selection. Academic and physical training were incorporated to prepare the astronauts, along with survival practices and flight control simulations. It wasn't until late 1959 when the Soviets had narrowed down their candidate pool to 20 that their training commenced. But due to the inefficiency of training all the cosmonauts, they cut most of them and selected six, a group that became known as the Vanguard Six – Yuri Gagarin (1934–68), Andriyan Nikolayev (1929–2004), Pavel Popovich (1930–2009), Gherman Titov (1935–2000), Anatoly Kartashov (1932–2005) and Valentin Varlamov (1934–80). The latter two were later replaced on medical grounds by Valery Bykovsky (1934–2019) and Grigory Nelyubov (1934–66). Cosmonaut training was gruelling, with a daily physical programme and classes covering rocket systems, astronomy, navigation and geophysics.

Although the Americans started their human spaceflight programme first and began training their astronauts ahead of their rivals, the Soviets had more automated spacecraft and could claw back time that would have been required for manual pilot training. By mid-1960, the USSR had become privy to information that suggested the Americans would be able to conduct a sub-orbital human spaceflight by the January of 1961. Fearful of defeat, the Soviets launched several uncrewed test missions during the latter half of 1960, which proved to be unsuccessful. By the time January 1961 rolled around, neither the US nor the USSR were ready for a crewed launch and each day became more critical – another day of much needed testing was a day that could cost them the race. By March, the Soviets had conducted two successful tests, the latter involved sending a dummy called Ivanovich and a dog named Zvezdochka in a prototype of the Vostok spacecraft, both of whom were recovered safely.

On the morning of 12 April 1961 at the Baikonur Cosmodrome, a similar yet heightened feeling of anxiousness and excitement, as when the Soviets had been preparing to make history with Sputnik 1, would have filled the air. For it was on this day that Yuri Gagarin became the first human to make an orbital spaceflight. It was only four days prior to launch that Gagarin was chosen for the spaceflight, though he had become a favourite among his peers well before. Gherman Titov was in the front running too but the head of cosmonaut training wanted to keep him (the stronger of the two) for the proposed longer one-day spaceflight. Prior to the flight, doctors noted Gagarin was unusually quiet and unsociable – no doubt the pressure and apprehension of the mission ahead would have been churning under the surface, and manifesting visibly through his otherwise composed exterior. Gagarin's heart rate in the half hour before launch was a mere 64 beats per minute – below the mean rate for an adult, and less than half of that experienced by the average person on a roller coaster anticipating the climax. Launched on board Vostok 1, Gagarin spent 108 minutes in flight and made an orbit of the Earth before returning. With no engines to slow the spacecraft upon re-entry, Gagarin had to eject and parachute to safety from a height of 7 kilometres. And although the FAI denote that a pilot must land with the spacecraft to count as an official spaceflight, Gagarin was without question the first person to travel into space.

News of his flight was announced while he was still in orbit, and upon return he became a national hero and an international celebrity. He was awarded the title of 'Hero of the Soviet Union', the highest honour from the USSR. The historical day was nationally declared Cosmonautics Day in commemoration, and in 2011 it became the International Day of Human Space Flight by the United Nations. Many countries expressed their congratulations officially, including the US who praised the USSR on their grand achievement. Many people remained with mixed feelings – a sense of pride for the accomplishment, but an underlying fear of the potential military ramifications for the ongoing Cold War.

Following the Vostok 1 spaceflight, President John F. Kennedy (1917–63) remarked that it would be some time before the US could match the launch technology of the Soviets. But the Americans weren't far behind and within a month Alan Shepherd became the first American in space when he made a suborbital flight in May 1961 – a flight lasting just over 15 minutes. Two months later Gus Grissom matched the feat and became the second American in space. A few weeks after the first successful Mercury flight, NASA announced its Apollo Program – a three-crew project that could take astronauts to a proposed space station, or take them around the far side of the Moon and back to Earth (circumlunar flight), or even land astronauts on the Moon.

Before the Americans were able to conquer a crewed orbital spaceflight, the Soviets once again asserted their dominance in the Space Race by launching Gherman Titov into a day-long orbital flight in August 1961. It was with John Glenn (the third Mercury astronaut to fly) that the Americans finally reached crewed orbital flight in February 1962 – he became the first American to circle the Earth and the fifth person in space. Three more Mercury orbital flights were made before Project Mercury ended in May 1963 (with Deke Slayton of the Mercury Seven being the only astronaut not to fly in the programme). The final Mercury flight piloted by Gordon Cooper matched Titov's achievement of a day-long spaceflight with the completion of 22 Earth orbits. The remaining flights of the Vostok programme were launched in pairs. The final flights of Vostok 5, which launched on 14 June 1963, and Vostok 6, which followed two days later, included another record. Vostok 6 was crewed by Valentina Tereshkova (1937–present) – the first woman in space.

It seems odd that in the male-dominated industry that space exploration began as, a woman was able to make such a notable stamp in the history books. In fact, Tereshkova had no intention of going into space – she worked in a textile mill and had a passion for parachuting. Her solo mission, in which she made 48

Earth orbits and spent almost three days in space, began indirectly in America. William Randolph Lovelace II (1907–65) was a flight surgeon who helped develop many of the tests for the male astronauts on Project Mercury, and he became interested in how women would perform in these same examinations. A private programme was designed to do just that, and Mercury 13 was the name given to the 13 women that passed the same tests as their male counterparts. But this programme was hidden from the public eye. Although they weren't part of NASA's astronaut programme, the director of USSR cosmonaut training had got wind that females were training to be astronauts, and believed it would be insulting to the patriotic Soviet women if the first woman in space was an American. It was this eagerness to avoid humiliation and Tereshkova's experience in skydiving that contributed to her selection as a cosmonaut. She remains the only woman to fly on a solo space mission, and became the first and youngest woman in space, aged 26.

With Project Mercury completed, attentions turned to Project Apollo. The Soviets had continued to make records and the prestige and political advantage of the US was being questioned. When President Kennedy came into office in 1961, he was convinced that America needed to assert its space superiority. With the help of his Vice President Lyndon B. Johnson (1908–73) who consulted with NASA, they set the goalposts for Project Apollo. The potential aims of the project were so far beyond what had already been achieved – the bar was being raised to a new height and no one knew for certain if it would even be possible. After evaluating their initial ambitions, it was deemed that it would be highly unlikely that the Americans could launch a space station before the Soviets, and there was uncertainty in whether they could reach crewed Moon orbit before them either. The best option, the most extravagant one, and the one with the greatest expense ($22 billion), was the one they committed to. With Project Apollo, the US committed to landing an astronaut on the Moon.

Preparing for Crewed Moon Missions

Having set the aims for Project Apollo, NASA's Space Task Group (the group of engineers that managed NASA's human spaceflight programmes) had to expand and Houston, Texas, was selected as the new site. Now known as the Johnson Space Center, it is where human spaceflight training, research, and flight control are conducted from in America. In September 1962, Kennedy visited Houston to view the new facility and it was at this time that he took the opportunity to get some backing for the nation's space efforts. Kennedy delivered his famous speech, entitled The Address at Rice University on the nation's space effort, where he declared with such passion:

> We choose to go to the Moon... We choose to go to the Moon in this decade and do the other things, not because they are easy, but because they are hard; because that goal will serve to organise and measure the best of our energies and skills, because that challenge is one that we are willing to accept, one we are unwilling to postpone, and one we intend to win, and the others, too.

Kennedy's speech was moving but many people weren't in support of the space effort at that time. People were disheartened by the nation continuing to fall short behind the USSR, and the huge costs involved that could have been invested elsewhere, not to mention the unrest from the civil rights movement taking place at the time.

Kennedy on several occasions did propose a joint expedition to the Moon with the Soviets, which was declined with the explanation that they had no intention of sending cosmonauts to the Moon. Even though military advisors of Nikita Khrushchev (1894–1971), the leader of the Soviet Union at the time, advised him to revaluate and take up the offer, the proposition of a joint mission was abandoned after Kennedy's death in 1963.

In order to develop the required technologies to make the Moon landing a possibility, Project Gemini (NASA's second human-spaceflight programme) was conducted from 1961 to 1966. This bridging programme would fill the gap between what had already been achieved with Project Mercury and the task they had now been set with the Apollo Program. The Gemini spacecraft could carry two astronauts, and between 1965 and 1966 there were 16 individual astronauts who flew in ten LEO missions. It was the first programme to make use of the newly built facilities in Houston. The corps included three of the Mercury Seven astronauts (Cooper, Grissom and Schirra), along with eight of NASA's second astronaut training group which included Neil Armstrong (1930–2012), and five astronauts from NASA's third group. Mission endurance, extravehicular activity (EVA), and orbital manoeuvres were tested during the Gemini missions, leaving Apollo free to concentrate on its prime mission. The astronaut-carrying Gemini spacecraft was the first to have an on-board computer to help with mission manoeuvres. Far from the computers we're familiar with today, the simple Gemini spacecraft computer was only able to perform calculations for orbital insertion or re-entry manoeuvres, after which the pilots would have to manually adjust the spacecraft's velocity appropriately. Provided they could successfully land on the Moon, a rendezvous in lunar orbit and docking would then have to be carried out, to bring astronauts back from the surface of the Moon to a command module, which would then bring them back to Earth. Getting two spacecraft to meet in orbit is incredibly tricky. Alike to when you swing faster on a roundabout and get flung outwards, when a spacecraft increases its speed to try and catch up with another it actually gets flung out to a higher and more distant orbit. Just as we see with planets orbiting the Sun, the larger the orbit, the slower the planet travels. So, a spacecraft that ends up in a higher orbit slows down and the distance between the two spacecrafts increases! Perhaps counterintuitively, a spacecraft should first position itself in a lower orbit where it will move faster and therefore catch up with the spacecraft it's trying to rendezvous with, before altering its orbital speed to approach it from below. Astronauts spent plenty of time in simulators practising complicated manoeuvres to master the technique.

Gemini 12 was the last mission in the programme (the first two were uncrewed) and in 1966, it drew to a close, having successfully demonstrated two things: that astronauts could work outside a spacecraft (through EVA) and that rendezvous and docking in orbit were possible. These accomplishments were instrumental in preparing for a crewed lunar landing before the end of the 1960s.

While Project Gemini focused on manoeuvring of crewed spacecraft and EVA, in parallel during the early 1960s, the Apollo program began to turn its attention to the launch vehicle. The Apollo missions would be three-crew space missions and the launch vehicle would need to have enough thrust to exceed the

Earth's escape velocity, while carrying a larger payload in order to reach the Moon.

For the five uncrewed tests of the Apollo spacecraft's launch escape system (LES), the smaller Little Joe II rocket was used. LES is a crew-safety system used in the case of an emergency, such as an impending explosion, to abort and separate the crew-carrying capsule from the launch vehicle. For all following flights in the Apollo Program the Saturn family of rockets were implemented. Saturn I (the first American heavy-lift rocket) was used in uncrewed suborbital and LEO flights to test hardware developments. The Saturn IB was the next iteration, with a more powerful second stage that would allow heavier payloads to be sent into LEO. This meant the Apollo spacecraft could be tested in space around our planet. But the final multistage Saturn V rocket was needed to reach the Moon, and the development of it was headed once again by von Braun. As

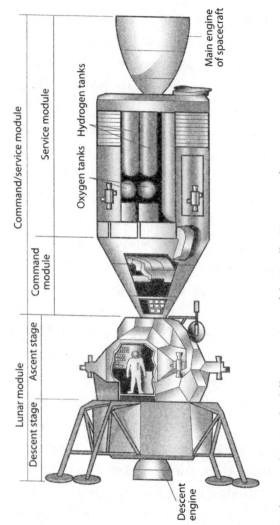

Main engine of spacecraft

Command/service module

Service module

Command module

Hydrogen tanks

Oxygen tanks

Lunar module

Ascent stage

Descent stage

Descent engine

A diagram showing the different sections of the Apollo 13 spacecraft.

of the start of 2021, the Saturn V remains the biggest and most powerful rocket brought into operation. Towering at 111 metres with the Apollo spacecraft atop, and weighing in at 2.9 million kilograms when fully fuelled, this record-breaking launch vehicle was designed to send a payload of 41,000 kilograms to the Moon – equivalent of lifting three unladen double decker buses into space and then some!

The Apollo spacecraft consisted of three parts: the command module, service module and the lunar module. The command module was the cabin for the crew and would be the only part that returned to the Earth, whereas the service module was the support system, like a hospital bed and life support system needed for a hospitalised patient. The lunar module had two parts – a descent stage to land astronauts on the Moon, and an ascent stage to get them back up into lunar orbit where it would rendezvous with the command and service modules, before journeying back to Earth.

The first test flight of the larger Saturn IB launch vehicle was conducted in February 1966, with two further uncrewed Apollo missions that tested the effect of weightlessness on the fuel tanks, and a suborbital flight of the command and service module. These missions were referred to as AS-201, AS-203 and AS-202 respectively, and initially considered Apollo 1, 2 and 3 but were later not numbered. NASA was getting ready to launch its first crewed mission AS-204 – a LEO test of the Apollo command and service module with commanding pilot Gus Grissom, command module or senior pilot Ed White (1930–67), and lunar module pilot Roger Chaffee (1935–67). But delays and testing failures meant that the first crewed mission was pushed back to February 1967, and even then some had expressed their concerns about certain safety aspects being overlooked. But with the success of the Mercury and Gemini Programs (where there had been no casualties in crewed spaceflights), perhaps a sense of complacency had taken hold. What would have been a momentous day with the launch of the first crewed Apollo mission, ended horrifically a few weeks prior to the planned launch. Throughout the launch rehearsal test in late January, there were glitches in communication between the spacecraft and Mission Control in Houston, leading to delays and forcing the test to run hours behind schedule. By early evening, still testing from within their small cabin perched at the top of the rocket, Grissom was frustratedly heard stating, 'How are we gonna get to the Moon if we can't talk between two or three buildings?' Not more than a minute later, engineers saw a flash on their screen – a spark had ignited a fire in the cabin and devastatingly killed all three crew members. Review investigations found the fire was ignited by a fault in the electrical wiring, and due to the combustible nylon material and the high-pressure cabin atmosphere of pure oxygen, the fire spread rapidly. The high internal pressure of the cabin to ensure the door shut tightly meant that a rescue attempt from the outside was prevented for several minutes, by which time it was too late to save the crew. Emergency preparedness was lacking, and harsh lessons were learnt from this terrible event. Even though the test was considered low hazard since the rocket was unfuelled, it was a brutal reminder that no corner could be cut or overlooked when it came to spaceflight. Upon request, NASA later officially named the mission Apollo 1 in the crew's honour.

Following the disaster, NASA continued to press ahead with the programme but made thousands of changes to the spacecraft; changes that could have saved the crew of Apollo 1. It was perhaps the wake-up call that was needed. NASA knew that there were problems with the programme but seemed to work around them to keep on schedule. The accident had a profound effect on everyone working on Apollo, and the more rigorous testing that followed meant that 20 months elapsed before NASA was ready for its next crewed Apollo mission. Between November 1967 and April of the following year, Apollo 4, 5, and 6 went ahead – these were uncrewed missions testing the Saturn IB and also the Saturn V rocket.

In October 1968, Apollo 7 launched carrying Donn F. Eisele (1930–87), Walter Cunningham (1932–present), Wally Schirra, in what became the Apollo program's first successful crewed mission to space. They fulfilled what the Apollo 1 mission was unable to do and tested the command and service module in LEO, circling the Earth 163 times and spending 10 days and 20 hours in space in the process. This was the first crewed mission since the final Gemini mission in 1966. But the end of the decade was fast approaching and there was still a long way to go in order to achieve the goal of reaching and landing on the Moon. However, the technical success of Apollo 7 gave NASA the confidence to make the next step – a crewed flight to the Moon.

Human Exploration
of the Moon

When thinking about the exploration of our celestial neighbour, people immediately recall the iconic first steps on the Moon, but another seminal mission paved the way for humanity's success in lunar exploration. Apollo 8 launched in 1968, just two months after Apollo 7, and with it, Frank Borman (1928–present), James Lovell (1928–present) and William Anders (1933–present) became the first astronauts to travel to the Moon. They escaped the gravity of a celestial body as they ventured through empty space towards our lunar neighbour. The spacecraft took almost three days to reach the Moon, and the crew orbited it ten times over a 20-hour period before returning back home to Earth. The famous photo *Earthrise*, taken by Anders from lunar orbit, is often considered the most influential environmental photograph ever taken. In it we see the Earth from beyond its sanctuary; we see its magnificence and its insignificance against the blackness of space, and these three astronauts were the first ever humans to witness it like this. No one had ever left the vicinity of the Earth prior to this. These astronauts travelled through space to a world that had never been visited before by humans. Not only did they and Mission Control have to precisely plan and carry out manoeuvres that would ensure they were captured by the Moon's gravity, (otherwise they would have ended up travelling further out and become lost in space), but they would also have needed to adjust their velocity at just the right time to ensure they went into lunar orbit. This adjustment had to take place when they were on the far side of the Moon, with no contact with anyone on Earth.

In low Earth orbit, Apollo 9 tested many critical aspects for landing on the Moon and was the first space test of the complete Apollo spacecraft – all three parts. Launched in March 1969 and spending ten days circling the Earth, the crew performed the first crewed flight of the lunar module along with docking and extraction of it. They demonstrated that it would be worthy of crewed spaceflight. With many of the individual aspects of a crewed lunar-landing mission tested, two months later Apollo 10 became a dress rehearsal for all the procedures and components that would be required for landing on the Moon – minus actually landing. It was the first mission to travel to the Moon with the full gear in tow; consisting of the command and service module dubbed Charlie Brown, and the lunar module called Snoopy. Crew members chose the call names for the command and lunar modules – a tradition which first began with Apollo 9 when the complete Apollo spacecraft began to be tested. After Apollo 10, NASA wanted to have more 'dignified' call names – some future command and lunar modules did, but Casper (the Friendly Ghost), the name chosen for the command module of Apollo 16, proves the astronauts still found a way to have fun with it.

Although the Soviets had stated that they did not intend to send cosmonauts to the Moon, they had not completely given up on maintaining their lead in the Space Race. They had been working on the N1 launcher – a heavy-lift rocket comparable to the Saturn V that could take payloads beyond LEO, but they were set back by repeated failures in its development.

Nevertheless, even though the Soviets may have accepted they could not beat the Americans to crewed landing on the Moon, they tried to triumph over the US in being the first to return a lunar sample to the Earth using robotic spacecraft. On 13 July 1969, the USSR launched Luna 15 – their second attempt to retrieve lunar samples.

This was just three days prior to the launch of Apollo 11 – NASA's fifth crewed Apollo mission and the one in which they intended to land men on the Moon. While the competition brewing from the Cold War would have burned away in the background, the safe launch of Apollo 11 on 16 July 1969 would have been what dominated NASA's attention. Parts of the spacecraft and launch vehicle had been arriving at the Kennedy Space Center (NASA's primary launch centre) since January of that year. This historic flight was a long time in the making –

more than the six months that this launch was being prepared, and more than the eight years that the Apollo program had been working towards this goal. This was the culmination of thousands of years of inquisitiveness from people who had gazed at the Moon and wondered what it was really like. When Neil Armstrong, Buzz Aldrin (1930–present), and Michael Collins (1930–2021) set off on their mission, sat in their cabin over 100 metres above the ground, on top of a fully fuelled launch vehicle with the energy of a nuclear bomb, the sense of excitement and anxiety in anticipation of the launch would have been subdued by the incredible focus that was required of them. It probably did not cross their minds how different their lives would be when they returned.

The astronauts would have experienced a g-force of over 3 g for close to one minute on their ascent, reaching a maximum of 4 g during launch. To put that into perspective, a person might experience just over 1 g during an emergency stop or sharp brake in their car, but that g-force only lasts for a short period of time, compared to the prolonged forces experienced by the astronauts. After 12 minutes of flight, they had entered Earth orbit and following one-and-a-half revolutions around the Earth, the engines were fired to take them away from our planet on a trajectory towards the Moon.

By 20 July, they had entered orbit around the Moon and Armstrong and Aldrin moved into Eagle, their lunar module, in preparation for the final descent. After making their last checks, Eagle separated from the command and service module (Columbia) – they were now into unknown territory. Everything had been tested up until this point, but Collins was now alone in Columbia watching as his colleagues made the perilous descent down to the surface of the Moon.

As they edged ever closer to the lunar surface, Armstrong noticed that they were passing landmarks on the Moon a few seconds earlier than they were expecting to, and as a result they might land miles from their target location. When 1,800 metres above the surface, the lunar module guidance computer raised an error. Initially the astronauts were not sure what the alarm meant and at that point an abort scenario would have become an option. After 30 seconds of waiting for clarification, Mission Control in Houston confirmed that the alarm indicated that the on-board computer could not complete all the tasks and calculations it needed to in real time, and had to delay some – but the mission was still a go. Although more advanced than the Gemini spacecraft computer, the technology of the Apollo guidance computer would easily be outsmarted by modern smart phones. It had a 1.024-megahertz processor (how quickly it carried out tasks) with 4 kilobytes of RAM (memory), making this integral 32-kilogram computing machine 100,000 times slower and far bulkier than a modern computer. Yet at the time, it was the most sophisticated bit of tech around.

Armstrong had spent hours memorising landmarks from photos that had been taken by the Apollo 10 crew on their dry run, and was once again noticing they were a bit 'long' compared to where they should have been. As Aldrin requested altitude data from their guidance computer, once again an alarm signalled. The computer crashed and had to be restarted. As they got closer to the surface with the fear that the problems could get worse with the computer, some of the software programmers back at Houston had internally concluded that the mission could no longer be 'a go', in their eyes it was time to abort. This, they only revealed years later.

Eagle continued its descent and the computer continued to fail and then come back. Armstrong had noted that their landing site would be a boulder-strewn area, not somewhere that the Eagle could be landed safely. They would have to alter their course and Armstrong would have to take semi-automatic control. Getting ever closer to the surface Aldrin called out the navigation data while Armstrong focused on piloting. And then, only 33 metres from the surface, Armstrong was adamant that they needed to land at the next clear site, knowing that the propellent supply that they were using to control their path was dwindling. Over three hours after the Eagle had separated from Columbia, Armstrong and Aldrin touched down on the lunar surface with less than a minute of fuel remaining for powered flight. Following a terrifying and eventful descent, the relief and pride that must have been felt by the thousands of people working on Apollo

A V-2 rocket used by Germany in the Second World War. It was the first artificial object to reach space after a vertical test launch on 20 June 1944.
World History Archive/Alamy Stock Photo

Using the Saturn V launch vehicle, the Apollo 11 mission took off from Kennedy Space Center, Florida on 16 July 1969. This iconic mission sent humans to walk on the Moon for the very first time.
NASA

Cosmonaut Valentina Tereshkova was the first woman in space, crewing Vostok 6 in 1963 at the age of 26.
Pictorial Press Ltd/Alamy Stock Photo

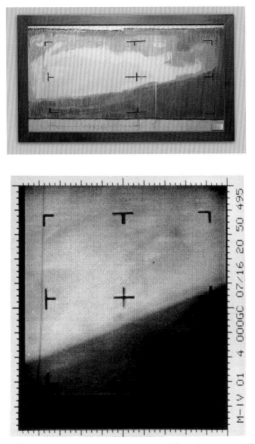

These are two representations (showing a good colouration match) of the first image returned from Mariner 4 from its flyby of Mars in July 1965. The top image is hand-coloured by scientists – they printed out the brightness values onto ticker tape, lined them up vertically and used pastel crayons to colour in the numbers. The other is the processed photographic image.
NASA/JPL-Caltech/Dan Goods

This image, entitled *Earthrise*, was taken from lunar orbit by astronaut William Anders on 24 December 1968 during the Apollo 8 mission. It was the first crewed spacecraft to leave low Earth orbit.
NASA

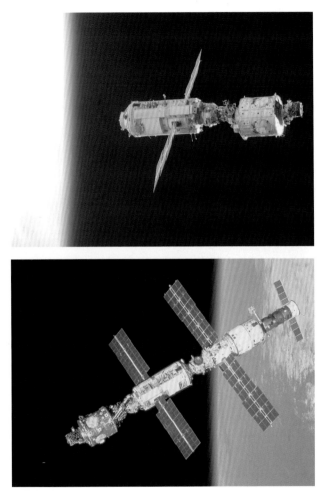

The International Space Station has mostly been assembled in orbit since its launch in November 1998. The top image, taken a month after launch, shows the mated Russian-built Zarya module (left) with the US-built Unity module (right). The bottom image from December 2000 shows how much it had grown in just two years.
NASA

Tim Peake took this selfie during a spacewalk on 15 January 2016. It took 4 hours and 43 minutes to replace a failed power regulator and install cabling on the International Space Station. *Geopix/NASA/Tim Peake*

Curiosity landed on Mars on 6 August 2012 and is one of several rovers that explore the planet, carry out scientific experiments and return footage to Earth. This 'selfie' was taken with a camera on the end of its robotic arm.
NASA/JPL-Caltech/MSSS

when Armstrong uttered the phrase 'the Eagle has landed', simply cannot be described in words.

It was over six and a half hours later that the astronauts were ready to open the hatch and step out onto the Moon, after checking the systems, donning their spacesuits, and preparing for lunar exploration. They spent little over two hours outside the spacecraft and within 22 hours of landing on the Moon, the pair were on their way back to rejoin Collins in lunar orbit.

In the limited time they spent on the Moon, they took photographs using the Hasselblad camera, collected 22 kilograms of lunar samples, and tested methods of walking around in the lower gravity environment (though Aldrin's two-footed kangaroo hops might be considered more for enjoyment than experimentation by some!) The slippery and fine nature of the lunar soil meant the astronauts had to plan six or seven steps ahead. They also planted the Lunar Flag Assembly, deployed scientific equipment to measure moonquakes, fitted a retroreflector array so that the distance to the Moon could be determined accurately by lasers, and even spoke to President Richard Nixon (1913–94) in the most historic telephone call perhaps ever made from the White House. What many often overlook is that during Armstrong and Aldrin's day-long venture, their co-pilot Collins remained in orbit around the Moon on board Columbia, solitarily. He would have been in communication with Houston at times, but during each of the passes he made behind the Moon he experienced the eeriest type of loneliness. At those times, with no communication with anyone on Earth or his colleagues on the Moon, he was the most isolated human in history.

After more than eight days in space, the crew returned to the Earth on 24

July 1969. They spent three weeks in quarantine before being given the 'all-clear' of a clean bill of health. Held in their honour, an estimated six million people attended the ticker-tape parades they rode in on 13 August in both New York and Chicago. Each astronaut was bestowed with the Presidential Medal of Freedom – a prestigious award from the US President to recognise their meritorious contribution. As expressed by the astronauts themselves, this was not a success for just NASA or the US; this was a success for the entire world and everyone living on it. Over 530 million people, 20 per cent of the world's population at the time, watched the Apollo 11 astronauts take the historic first steps on the Moon, and having returned to the Earth their accomplishment had proudly become one which the world wanted to celebrate with them. As such their achievements were commemorated in a 38-day world tour, taking them to 22 countries. President Nixon also gifted 270 Moon rocks to foreign countries as a token of good faith and a symbol of peace.

The Russian robotic spacecraft Luna 15 which had set off a few days prior to Apollo 11 did reach lunar orbit on 17 July while the crew of Apollo 11 were still on their way to the Moon. The Soviets spent a few days checking all of its on-board systems by which time the Apollo astronauts had landed. While they were finishing their spacewalk, Luna 15 did prepare to descend to the surface, but it unfortunately crashed just a few hours before the Apollo 11 crew were readying to return home. It effectively erased any hope of the Soviets returning lunar soil samples back to the Earth before the Americans. Even after the Apollo 11 mission, the Soviets maintained that they weren't interested in crewed lunar landing – it was simply dangerous and unnecessary, and instead they were focused on large satellite systems. However, two decades later it was revealed that the Soviets had been racing the Americans to land humans on the Moon, but numerous technological failures meant they were unable to compete. One huge disaster included the death of cosmonaut Vladimir Komarov (1927–67), when the parachute intended to safely settle the Soyuz capsule (in which he sat) back on Earth, failed to deploy. It was the first in-flight death of any person in space exploration. The perceived shame of their efforts and mounting problems with their launch vehicle is what turned the Soviets to instead focus on robotic missions to the Moon and Venus.

Although Kennedy was not around to witness it, his goal had been fulfilled – a man had landed on the Moon and returned safely to the Earth before the end of the decade. Apollo 11 essentially ended the Space Race, but further Apollo missions continued to press ahead as NASA had already planned for further lunar geological and astrophysical exploration. In the end, budget cuts meant that the last few Apollo flights were cancelled, making Eugene Cernan (1934–2017) of Apollo 17 the last man to step off the Moon in 1972.

Out of the six missions that followed Apollo 11, only Apollo 13 was prevented from landing. On route to the Moon, an oxygen tank explosion affected the electrical power supply severely crippling the propulsion and life-support system on the command and service module. Using the lunar module as a lifeboat, the crew looped around the Moon without landing and returned safely back to Earth. Apollo 13 encountered one problem after another. In trying to solve immediate problems utilising the limited resources at their disposal, they inevitably created other issues further down the line. They used the spacecraft and equipment in ways they were never intended to be used in order to survive. As they drifted back to Earth after swinging around the Moon, almost everything was switched off to save power for the descent back to Earth. This meant the astronauts were cramped together for several days in temperatures as cold as 10°C in the lunar module, and just 4°C in the command module that had been powered down. Thankfully, all three astronauts returned safely back home from their mission which became known as the 'successful failure'.

The five other missions conducted further scientific experiments including deploying another seismometer experiment to detect vibrations and tilting in the lunar surface, and measure gravity changes at its location. The astronauts took further photographs and collected more lunar samples – a total of 382 kilograms consisting of 2,200 separate samples were collected from the six different landing sites on the Moon in the duration of the

Apollo program. Pieces of the robotic Surveyor III lander that had been on the Moon for over two years prior to Apollo 12's mission were retrieved to study the effects of the lunar environment on them. For the last three missions, astronauts also had a lunar rover to drive around on the surface, which meant they were able to explore much further from their respective landing sites.

With all these great feats, the unfortunate casualties of such a fast-paced and perilous enterprise tend to go unremembered. Their sacrifices provide insights and a reality check to remind us of the serious dangers that spaceflight presents each and every time we embark on an expedition.

The deaths of three astronauts in air crashes during training in the earlier Gemini program along with the loss of the Soyuz 1 cosmonaut in spaceflight, and the three astronauts in the Apollo 1 cabin fire, serve to remind us that human spaceflight is perilous. Those dangers begin with the training here on the Earth, and are only magnified when it comes to the actual launch. But these risks then remain at every single moment that astronauts spend in space away from the comfort and safety of the Earth. Nevertheless, as Apollo 1 astronaut Gus Grissom put it, 'The conquest of space is worth the risk of life,' and so the development of technologies to allow humans to explore space further continued.

Space Stations
Through the Decades

Although during the 1960s both the Soviets and Americans had focused on lunar exploration, many programmes simmered in the background – projects that would allow each country to build on and maintain their lead in space exploration, once crewed lunar landing had been achieved. One such proposition conceived around 1967 by NASA members, with the idea of reusability and sustainability in mind, was to replace its expendable rockets with a reusable shuttle which could make repeated flights into low Earth orbit. Initially known as the Space Transportation System, this proposal for the successor to the Apollo program could be used long after the project to support future space programmes, like a permanent orbiting space station around the Earth, an outpost around the Moon, as well as a crewed mission to Mars. But having achieved the goal of landing a man on the Moon, funding for additional space exploration programmes was significantly disappearing and so a scaled-back version of the Space Shuttle program, as it then became known, was accepted but deferred in time, and the first shuttle only flew in 1981.

Over in the Soviet Union, Almaz (an extremely secretive USSR military space-station programme) was proposed as early as 1964, and this orbital piloted station was intended to operate for just a few years in which time it would take photographic and radar reconnaissance images. Almaz was conceived in response to the 1963 announcement of the United States Air Force's (USAF) programme, MOL – Manned Orbiting Laboratory. Even though USAF were open about its military nature for MOL to be an inhabited station to demonstrate the practicability of putting people in space for military missions, they did (like the Soviets) keep the reconnaissance element of it undisclosed. But MOL suffered from repeated budget cuts as funds were deemed more necessary for the Vietnam War, and as automated system technologies improved rapidly, a crewed or manually controlled space platform became regarded as less worthwhile. By 1969, MOL was cancelled before any crewed missions were even carried out.

During the 1960s, NASA had roughly 400,000 people working on the Apollo program and decided to put together the Apollo Applications Program (APP), which would find ways of modifying the hardware that was being built for Apollo for more varied scientific missions, to help provide job security after achieving the goal of landing on the Moon. Even though he had more ambitious plans that were rejected, von Braun advocated converting a Saturn V launch vehicle into a space station as part of this programme. Shrinking budgets and simplification of plans eventually led to the development of Skylab – an orbital space station converted from parts of the smaller Saturn I rocket.

Having been on the backburner for years, the Soviets made haste with building Almaz following the successful lunar landing in 1969 and the US announcement of Skylab. However, to conceal its military nature, Almaz became known as the Salyut Space Station, designated for civilian purposes. Salyut became the first space station programme with the launch of Salyut 1 in 1971. Over 15 years, the Soviets successfully launched six Salyut stations of which two were for crewed military reconnaissance. These stations operated in LEO with orbital heights between 200 kilometres and 280 kilometres above the Earth's surface. The purpose of the four-crewed scientific research space stations was to carry out long-term research into the problems of living in space, along with looking at a multitude of astronomical and biological experiments, and even Earth resources studies. Cylindrical in shape, these Salyut stations were no larger than 20 metres, with a diameter of roughly 4 metres. Salyut had a monolithic design, meaning that it was constructed and launched in one piece. Any crew would be launched separately to inhabit the space station but once the on-board supplies had been used up, the station would be abandoned until a new crew arrived. The Salyut flights once again raised the bar of achievements with various spacewalk and spaceflight records being set, including three cosmonauts who spent more than seven months in space on board Salyut 7 in 1984 – longer than anyone had previously inhabited space before. Each iteration of Salyut stations was an improvement on the last. The first-generation stations lasted no more than two years in orbit, as over time the minuscule drag force experienced from the tenuous atmosphere several hundred kilometres above the Earth's surface would send the spacecraft into a decaying orbit, at which point it would have to be deorbited. But the final two stations, Salyut 6 and 7, were second-generation stations and had two docking ports. This meant that refuelling to continue controlling the station's position was possible, and a second crew could visit bringing new supplies, allowing for a crew handover and for the station to remain continuously occupied. After almost nine years orbiting the Earth, before it was discarded, Salyut 7 underwent a hard (or permanent) docking as a development of the concept – it served as proof that space stations could be designed to be modular, that they could in fact be taken up in parts and built in space!

The 25-metre American Skylab space station launched in 1973, a couple of years after the first Salyut station. It too operated in LEO but at a height of 440 kilometres above the Earth's surface. Unlike the raft of Salyut stations, each one replacing its predecessor, Skylab was planned to last nine years, so was built to go up and not intended to come down before that. Crews would be sent up with supplies and return once they were depleted. But by 1978, Skylab's orbit was decaying, sending the station on a path that would bring it crashing back down to the Earth far earlier than expected. NASA's first response was to get the in-development Space Shuttle to boost Skylab back into a higher orbit, but delays in the project meant that NASA had to come up with a new plan and fast, as Skylab was descending from its orbit rapidly. In 1979, firing the station's booster rockets, NASA planned to send the 76-tonne Skylab on a trajectory that would bring it hurtling into the Indian Ocean. Disintegrating on its descent, most of it crashed into the ocean as planned, but parts of the space station were sent flying into areas of Western Australia. Fortunately, no one was injured, even though the litter fell in populated areas.

The second generation of the Soviets' Salyut programme paved the way for modular space stations. The core component that was initially developed for the next generation of Salyut stations evolved to become the central component of Mir – the first ever modular space station. This iconic Soviet station was assembled in space beginning in 1986 and took a decade to complete. It orbited

360 kilometres above the Earth and the core module had six docking ports allowing spacecraft to visit and additional modules to be hard docked or attached. 'Mir' translated as 'peace' or 'world' and was later operated by Russia after the fall of the USSR in 1991. With additional modules attached, it was the largest artificial satellite of its generation in orbit, spanning 31 metres across. In LEO, the station was a microgravity research lab – a place for crew to conduct all sorts of science experiments, and test engineering technologies that would allow humans to permanently occupy space. Out of its 15 years in operation, Mir was inhabited for 12 and a half of those, with a typical resident crew of three, but allowing for larger crews on short visits.

Following the dissolution of the Soviet Union and the establishment of Roscosmos State Corporation for Space Activities (more well known as Roscosmos), the

Shuttle–Mir program blossomed. This collaboration between Russia and the United States was developed to encourage a spirit of cooperation between the two nations, allowing each to learn from and experience the other's technologies and insights – a partnership which lay the foundations for the future construction of a truly international space platform. Between 1993 and 1998, American astronauts were launched aboard Russian Soyuz spacecrafts (used to transport people to and from Mir), and cosmonauts flew in the US Space Shuttle. Long duration missions of American astronauts to Mir also occurred. They undertook training in operation of Mir and the Soyuz spacecrafts and were flown to Star City (a training facility) in Russia to practise spacewalks. They even had lessons in the Russian language to be able to communicate with the cosmonauts on Mir and Mission Control in Russia – a training practice that is still in place today. Many nations took part in cooperative programmes too and the UK set up the privately funded campaign Project Juno to claim a seat for a Briton on a Soyuz spacecraft that would take them to Mir. It was Helen Sharman (1963–present), a chemist and food technologist by trade, who was selected for that mission, and in 1991 she became the first British cosmonaut, the first woman to visit the Mir Space station, and the first Western European woman in space.

In 1993 the US and Russia agreed to merge their separate plans for space stations to create a single facility that would incorporate their already designed modules and would also include contributions from ESA (European Space Agency) and JAXA (Japan Aerospace Exploration Agency). Construction of the ISS began in 1998 with the launch of the Russian control module Zarya. Later, the United States' Unity module was connected to it in orbit by Space Shuttle astronauts. The Russian-built Zvezda module (an evolution of the Salyut station) was used in the ISS as a habitat and control centre. Although the ISS took a decade to construct in its entirety, preparations began for the ISS to receive its first crew following the addition of the Zvezda module in mid-2000. Since Russian cosmonauts Sergey Krikalyov (1958–present) and Yuri Gidzenko (1962–present) and American astronaut William Shepherd (1949–present) arrived on 2 November of that year, the ISS has remained a continuously occupied outpost in space orbiting 418 kilometres above our heads! At almost 110 metres across (about the size of a football field), the ISS has claimed the prestige of being the largest artificial satellite in orbit since Mir's orbit decayed and it was brought down in 2001. With 16 modules, the interior is described as having the space of a five-bedroom house (albeit in a very tube-like configuration), where astronauts generally spend six months on board. Since 2009, the ISS has had the capability of comfortably housing six crew members. However, seven crew members began living on the ISS for the very first time in November 2020, after four new astronauts launched to the ISS, becoming the first commercial crew to join the existing inhabitants. The limelight of this occasion was rather humorously stolen by a Baby Yoda toy which wasn't just there as a cosy companion. It's been a long-standing tradition of spaceflight that the crew choose a toy to join them on their launch serving as a gravity indicator. Once the toy begins to float, the crew know they're in space!

The ISS is now a collaboration between five space agencies including the Canadian Space Agency (CSA) and accounts for almost 50 per cent of all the people who've been into space, having hosted over 240 people from 19 different countries in its first 20 years of operation. Although other UK-born astronauts have boarded the ISS, when Tim Peake (1972–present) made his maiden flight in December 2015, he became the first (and to date the only) British astronaut bearing the flag of the United Kingdom to inhabit the ISS, having surpassed 8,000 other applicants to take one of six places on ESA's training programme in 2009.

From the Mir Space Station, its monolithic predecessors, and in the 20+ years that the ISS has been operating in space, there has been a wealth of lessons and information we have acquired about living in space.

1. One of the biggest threats to health while in space is loss of bone mass and muscle atrophy due to microgravity. Other effects due to lack of gravity like

weakening heart muscles also need to be monitored. Our bones and muscles remain strong responding to the stresses they are put under by gravity. Research from the Mir Space Station showed that about 1–2 per cent of a person's bone mass is lost for every month they spend in the microgravity environment of LEO, where the pressure on our bones is significantly reduced. It's why there is a regimental two-hour exercise period scheduled into an astronaut's working day and even then, astronauts returning from a long-duration spaceflight have to be carried out of their spacecraft after returning to the Earth. It can take hours or even days for muscles to readjust and for walking to feel normal again, though it can take months to regain the lost bone mass.

2. Psychological health and wellbeing are incredibly important too, especially

in the confined regions of a space station with the same crew and a lack of orientation. So designs and certain social considerations for future crewed space exploration missions will need to be addressed to maintain an effective social habitat. Skylab had very little colour so the astronauts would stare at the coloured cards which were used to set up their video cameras in order to help break the boredom. In the Russian modules of the ISS, surfaces facing towards the Earth ('down') are coloured green to help give a sense of orientation. Unused surfaces are used like fridge doors, as astronauts have taken to sticking images and personal paraphernalia to express themselves and connect them to home. Sharing food, celebrating birthdays, and enjoying different festivals helps to build camaraderie within a crew too. On a long journey into space with no quick return home, a crew needs to work well together and remain socially content to deal with the challenges that human spaceflight inevitably entails.

3. Food can be resupplied on cargo ships but preparing and eating food in a microgravity environment comes with a limited menu. Even though the ISS may be depicted as a super clean place, dead skin cells, dust, and even crumbs from food which would normally settle on the floor, remain air bound instead. Cleaning is therefore another routine activity that takes place. To minimise the mess, ingredients are taped down when preparing foods, and meal trays are strapped to astronauts when eating to prevent things from floating away. Long-life foods are prepared and sent up – these are often rehydratable foods. Biscuits and bread are a no-go due to the crumbs, and even salt and pepper

come in liquid form to prevent sprinkles from flying off. For longer missions into space, astronauts will need to take all the food they need with them, or grow it themselves – resupply cargo won't be an option. So far, wheat and rice have been grown in space along with leafy greens like lettuce and cabbage, and even root vegetables like onions, radishes and garlic. But there's a long way to go to make this sustainable.

4. The invisible threat of radiation is another risk to health. On the Earth, the atmosphere and our planet's magnetic field protect us from a large amount of cosmic radiation. Radiation levels increase as you leave the planet's protective shields and so on the ISS which actually passes through the inner Van Allen Belt (a radiation zone around the Earth) on part of its orbit, radiation levels can be 30 times greater than on

the Earth. Within a week, astronauts can be exposed to a year's worth of radiation as they would on Earth. Since the Apollo missions, astronauts have reported seeing flashes of light, even with their eyes shut. Thought to be caused by cosmic rays triggering a response on the eye's retina, the effect is still being investigated today with an experiment on the ISS conducted by ESA astronauts called ALTEA (Anomalous Long-Term Effects in Astronauts). If we hope to venture further into space where the protection of the Earth will further dwindle, understanding the long-term neurological effects will be paramount. This means that astronauts have themselves become experimental testbeds. In fact, astronaut Scott Kelly spent a year in space beginning in March 2015 and any changes in his physical, neurological, biological and cognitive state have been and (for the long-term impacts) continue to be investigated with his twin brother Mark, who remained on Earth, acting as a 'control' subject for comparison.

5. Low Earth orbit is the ideal environment to begin investigating the effects of living in space. Still in the vicinity of Earth, resupply missions aren't out of the question. But even so, when things break or need repair, astronauts are sometimes employed to fix them. On occasion, fixes can require astronauts to leave the protection of the ISS and head outside, with a spacesuit being the only thing between them and inhospitable space. The first spacewalk was conducted by Alexei Leonov (1934–2019) in 1965 and even though many have performed the same feat since, most if not all recount it as being the most exhilarating yet unnerving experience they've had. With limited

dexterity in the bulky spacesuits and just a tether as your foremost safety measure, spacewalks are exacting work. With no containment to act as a safety net, a mistake could leave an astronaut floating off into endless space. A momentary lapse of concentration has meant that a mirror, cloth and even a tool bag have encountered that fate! As we prepare to explore space further, working in the exposed vacuum of space is a challenge we'll need to master.

Since its inception, NASA estimate over 2,500 experiments have taken place on the ISS. This incredible feat of human engineering is intended to operate up until 2030, but it won't then be left to circle the Earth as a derelict satellite. It will instead be brought back to the Earth like many decommissioned space stations through a controlled re-entry. Most likely, it would head for the 'spacecraft cemetery' – the final resting place for many spacecrafts and satellites including Mir. This burial ground is located in the South Pacific Ocean, an area that lies farthest from any landmass, making it the safest place to intentionally crash-land a spacecraft.

Tiangong was a space-station programme operated by the China National Space Administration (CNSA). Smaller than the Salyut space stations, the 10.4-metre Tiangong-1 space station could support a crew of three taikonauts (Chinese astronauts). It was launched in September 2011, and eventually in April 2018 it was deorbited and re-entered the Earth. This single module prototype space station was CNSA's first step in developing a modular space station comparable to Mir. Two more iterations were planned but with delays to the programme, the goals for Tiangong-3 were merged into Tiangong-2. Tiangong-2 served as a test bed to investigate integral technologies, so it was deorbited as planned in 2019, almost three years after its launch. The Chinese large modular space station is the next step and will manifest into the Tiangong programme's third phase, but the planned launch of its core module in 2020 was delayed.

An event that has blemished the success of this programme was the uncontrolled re-entry of the Tiangong-1 space station. After its orbit began decaying, **telemetry** failures meant that the Chinese controllers were unable to plan a precise engine burn that would direct the station to a designated safe area on Earth like the spacecraft cemetery. Only in the final hours of its descent were they able to track its path, and thankfully it plummeted into the expanse of the Pacific. Many space agencies expressed criticism exclaiming that with the growth in space exploration, we should understand that the responsibilities of a space mission encapsulate the entire cycle: from deployment, followed by operation as well as retirement, including safe disposal.

With our huge leaps in space exploration, we need to be equally focused and attentive to our wider responsibilities of sustainability and preservation. With over 60 years of space activity and thousands of launches in that time, there are numerous items orbiting the Earth and very few of them are operational satellites. Space debris is becoming a very real concern – even the tiniest fragments can travel around the Earth at several thousand kilometres per hour and striking a satellite like the ISS can cause serious damage. Not only are we having to track the objects orbiting above us, but we have the problem of needing to clear up this region of derelict objects. Some have proposed deorbiting missions using a capture mechanism like a huge space net, while other ideas include a solar sail or electrodynamic tether that will force space debris into lower orbits, where they would eventually burn up.

As we look to the future, the idea of space stations has expanded beyond Earth orbit. As part of NASA's Artemis program to return humans to the Moon by 2024, and prepare for a future crewed spaceflight to Mars, an orbiting space station around the Moon known as the Lunar Gateway has been planned. Serving as a solar-powered hub for communication, this minimalistic, modestly sized station will also act as a science laboratory, a place for humans to reside for short periods, and a space where rovers and other robots can be held. Ever-changing plans, delays and costs mean that there's uncertainty in whether the Lunar Gateway will be ready to support the planned return to the Moon in 2024 as initially proposed, but such a platform will most likely be pivotal if we want to establish long-term exploration and settlement, by acting as a waypoint for future missions to the Moon and further.

Sustained human space travel may take decades to develop, not only in terms of the technological advancements and engineering feats, but also for us to build a good foundation of knowledge on the effects of living in space and how we might be able to counteract the negative impacts. While we've spent decades investing in human spaceflight, we've also continued to send uncrewed robotic spacecraft to the far reaches of our solar system. We've also constructed space telescopes to allow us to study objects far beyond their vicinity, so that when we are eventually prepared to travel further out into space, we have a wealth of information to guide us, rather than stabbing in the dark and hoping for the best. Space simply isn't a forgiving place.

The Future of Space Missions

For the very reason that space exploration is a tricky endeavour, uncrewed robotic spacecrafts (the less risky option) have always been proposed, developed and launched even when our focus had been on crewed missions as during the latter part of the Cold War. They are the most viable way of exploring space. We often trivialise just how monumentally big our solar system is, let alone the Universe in which it sits. Although smaller robotic spacecraft might be able to travel faster, it can still take years to journey within our solar neighbourhood to reach other celestial bodies, which we imagine as being relatively close. There have been numerous space missions since the launch of Sputnik 1, only a few of which are outlined in this chapter along with glimpses into proposed missions that professional astronomers and those with a keen interest await with great excitement and anticipation.

The Moon is undoubtedly the most well-explored extraterrestrial body in the Solar System. Notable missions like the Soviet Luna probes and the multiple NASA programmes like Pioneer, Surveyor, and Lunar Orbiter were among the first spacecrafts to explore the Moon beginning in the late 1950s and into the 1960s. Before the first crewed mission to the Moon in 1968 by the astronauts of Apollo 8, numerous robotic spacecraft had already flown by and impacted its surface. They had even entered lunar orbit and soft-landed on the Moon to photograph, test, observe, and map the lunar surface, giving us an up-close view of our nearest neighbour as we'd never seen it before.

As early as 1962, spacecrafts had reached another planetary body. In the midst of the Cold War, while the Soviets and Americans were racing to get the first person in orbit around the Earth, other groups within the space corporations were developing missions to explore Venus and Mars. NASA's Mariner 2 became the first spacecraft to successfully encounter another planet when it flew within 35,000 kilometres of Venus. Although that may seem quite far, it's pretty close when you consider that Venus is on average 41.4 million kilometres away from Earth. Compounded with numerous challenges, Venus is a notoriously difficult world to explore. One might assume that travelling into the inner Solar System would be easier – the Sun's gravity would provide a free propellant, reducing feel costs or the need for giant solar panels. But the opposite is in fact the case. Often more fuel can be required for missions to Mercury and Venus to overcome the enormous pull of the Sun. Probes need to be slowed down to be captured into orbit when they approach the inner planets. Getting that wrong would send a spacecraft careering past its planetary destination and perhaps into the burning hot clutches of the Sun, or it could end up nose-diving into the planet itself. It wasn't until 1975 that the Soviet Venera 9 probe successfully entered Venusian orbit. A few years before that, the Soviets had managed to successfully deploy atmospheric probes and even land on the surface of Venus.

From above, Venus' thick clouds obscure any view of its surface – they reflect about 70 per cent of the Sun's light and make Venus appear as a bright star in our twilight skies on Earth. Its beige-yellow hue results from the high levels of sulphur dioxide within them. The clouds also feature lightning storms and sulphuric-acid rain. The Venusian atmosphere, made almost entirely of carbon dioxide, is so thick that the air pressure on its surface would make it feel as if you were 1 kilometre underwater here on Earth. This suffocating blanket traps the Sun's heat bringing surface temperatures to a whopping 470°C – hot enough to melt lead. Considering the abominable conditions, it's surprising that spacecraft have endured the descent, landed, and operated on its surface, albeit not for very long.

It was Venera 8 that became the first probe to victoriously land on another planet in 1972. It lasted only 50 minutes due to the scorching temperatures and crushing atmospheric conditions. Venera 9 (the 1975 orbiter) also consisted of a lander which became the first spacecraft to take images from the surface of another planet. It too met its harsh fate, succumbing to the excruciating environment within an hour. Thankfully, the lander had two cameras so even though one lens cap didn't release, the other was still able to acquire images. Venera 10 also returned images from the surface.

Launched from the Space Shuttle in 1989, the Magellan spacecraft mapped Venus' surface from orbit using radar to probe through the otherwise impenetrable clouds. And using the colour images provided courtesy of the Venera landers, the radar data could be turned into near-realistic global views of Venus. Although the surface of Venus is barren and inhospitable, higher up in the toxic clouds of Venus it is more temperate. In the autumn of 2020, a team of astronomers (including the Royal Observatory Greenwich's own Dr Emily Drabek-Maunder) reported the detection of the chemical phosphine in the thick Venusian atmosphere. Phosphine can be created by biological processes – it's found in the faeces of badgers and penguins on Earth, for example. With no other known process that can explain the concentrations of phosphine found on Venus, it suggests the possibility of life in this most unimaginable place!

Mars is instead a place that has piqued our interest when it comes to alien life. This planet has fascinated humans since it was first observed. Its bright rusty colour, evoking images of conflict and bloodshed, led to the Romans naming it Mars after their god of war. With the invention of the telescope, dark lines appearing on the surface of Mars (a result of using low-resolution equipment) were termed *canali* by astronomer Giovanni Schiaparelli (1835–1910) in the late nineteenth century. The misinterpretation of these features as canals instigated mass excitement about the possibility of life on Mars – an engineering network of waterways surely was a sign of civilisation. Although the error was eventually recognised, our intrigue with Mars has never stopped.

In 1965, Mariner 4 made the first ever flyby of Mars, reaching about 9,800 kilometres at its closest approach, and captured the first images of the planet. Digital photography has come a long way in the past few decades and seems far-removed from the Mariner 4 image data that was captured on a tape recorder. Like pixels on a screen, markings could be captured on a reel of tape that were indicative of the light signal captured, which could then be processed into an image. Lining strips of tape side by side, scientists used pastels in a 'colour by number' fashion to create a hand-processed masterpiece. They assigned different shades of colour to the different markings and, to their delight, their hand-coloured image showed good resemblance to the digitally processed one.

From the early days of flybys and first images, Mars has now become the most explored planetary body in the Solar

System. A decade after Mariner 4's flyby, NASA placed a pair of identical landers on the surface of Mars, Viking 1 and 2. They also had their counterpart orbiters circling the red planet, which acted as communication relays once they'd touched down. Not only did they take photographs, but several biology experiments were conducted, setting into motion a research aim that is still prominent in Mars missions today – the search for signs of life. Although the Viking missions were planned to last for 90 days post-landing, the orbiters and landers operated for years, with the final signal coming from the Viking 1 lander which sent its last transmission in 1982.

Over the decades NASA has continued to dominate exploration on Mars thanks to missions such as:

- The Mars Pathfinder, consisting of the Sojourner rover which collected huge amounts of data exploring an ancient flood plain region. This first roving mission on Mars in 1997 changed the face of exploration on the red planet.

- The Mars Exploration twin rovers Spirit and Opportunity – these legacy rovers which landed on Mars in 2004 both discovered numerous signs to suggest that liquid water once flowed across the surface of Mars in the past.

- The Mars Reconnaissance Orbiter, with its telescopic camera (the most powerful camera sent to another planet) which helped map the Martian surface. In 2014 it confirmed that the British-built Beagle 2 probe of 2003 had actually landed intact but one of the solar panel mounts didn't unfold, which prevented the antenna from sending a message home.

- Insight – the first mission designed to explore the deep interior of Mars. Its seismometer detected the first Mars quake in 2019 and has since detected many small quakes, none stronger than magnitude 4, which has left scientists with a puzzle. Is Mars less active than anticipated or is Mars just going through a particularly quiet phase?

- Mars Science Laboratory, better known as the Curiosity rover, found mudstone in Gale crater where it landed in 2012, indicating that a large standing lake must have once existed there. It even discovered evidence to suggest that billions of years ago when it was likely a warmer place, Mars likely had a more oxygen-rich atmosphere.

- The Mars 2020 rover Perseverance will be scouring a delta plain deemed a habitable environment within the Jezero crater. It will be looking for signs of life but also caching samples for future return, and also testing if oxygen can be produced from the thin atmosphere. Furthermore, it has a robotic helicopter named Ingenuity, a technology test which in April 2021 made history when it demonstrated the first successful powered flight on another planet – hovering as well as travelling across the Martian surface in subsequent test flights. This triumph of engineering will pave the way for rotorcrafts to be used to seek out further places of interest in the future Mars missions.

There have been over 50 missions to Mars by a number of countries and many more planned, like ESA's Rosalind Franklin rover set to launch in 2022. Yet the success rate is only around 50 per cent, showing just how difficult a mission to one of our closest planetary neighbours really is. Much of the time it comes down to a software error, an engineering mistake, or just bad luck.

Despite the challenges, Mars is a target for a number of space agencies. As of the end of 2020, NASA remained the only country to successfully operate rovers on Mars, but with the arrival of the ambitious Tianwen-1 mission from CNSA, that has now changed. China asserted their increasing capabilities in space with a successful orbiter, lander and rover in this single Mars mission, making them the only country to achieve this mammoth feat on their inaugural mission, and the third to achieve a successful soft landing. After arriving at the red planet in February 2021 along with NASA's Perseverance rover and the United Arab Emirates' (UAE) Hope orbiter, the lander and rover component of this Chinese mission became the latest inhabitants on Mars when they touched down in May 2021. After a week of tests and checks, the six-wheeled solar-powered Zhurong rover drove down the landing platform and made China only the second country in history to roam on the red planet.

But to make human travel to Mars viable, we'll need to use the Moon as a stepping stone. Not only as a ground to develop our understanding of the effects of living and working in space, but perhaps also as a physical launch pad with our missions to Mars taking off from around the Moon instead of from Earth.

The Lunar Reconnaissance Orbiter launched by NASA in 2009 has imaged the Moon's surface providing a global map with 100-metre resolution of its landforms and features. In more recent times the Indian Chandrayaan missions have massively boosted its national space programme. Chandrayaan-1, which launched in 2008, discovered water on the Moon. With a thin atmosphere and weak gravity, water would evaporate and be lost into space. But at the poles of the Moon, in the permanently shadowed regions of craters, Chandrayaan-1 detected water, and its impactor probe also found water molecules in the very thin atmosphere on its descent. Chandrayaan-2 consisted of an orbiter (which went into lunar orbit in 2019) but communication with its lander and rover component was lost on descent, and it failed to touch down softly. Chandraayan-3 is scheduled for launch in 2022 – with only a lander and rover, it will attempt to succeed where the Chandraayan-2 mission failed.

China began its Chang'e project on lunar exploration in 2007 with the launch of Chang'e 1. The first phase of this endeavour would focus on orbiting lunar missions. A second orbiter, Chang'e 2, followed in 2010. Both probes are now out of operation but following their success, phase 2 of testing the landing and roving technology on the Moon was implemented. When Chang'e 3 landed on the Moon in 2013, it became the first to soft-land since the Soviet's Luna 24 mission in 1976. Chang'e 4 replicated the success of its predecessor in 2019 with one huge difference – it soft-landed on the far side of the Moon; a feat never before achieved. Phase 3 set the bar higher still. Chang'e 5 and 6 would bring back samples from the near side of the Moon. In late 2020, Chang'e 5 set off for the Moon. It collected roughly 1.7 kilograms of lunar samples, including some from 1 metre beneath the surface, and returned to the Earth on 16 December 2020 with the first lunar samples fetched since the 1970s. Having targeted a region called Mons Rümker, believed to have formed about 1.2 billion years ago, the Chang'e 5 samples are likely to be younger than the samples returned by the Apollo and Luna missions, and in time should provide more information to help us understand the Moon's internal structure. Chang'e 6 is planned for launch around 2024, and beyond that the Chinese Lunar Exploration programme aims to transition to phase 4 – the development of a robotic research station towards the south pole of the Moon. The elaborate extension is to pave the way for crewed lunar landing into the 2030s, with the possibility of an outpost for longer-term habitation.

NASA aims to achieve crewed lunar landing once again by 2024 with its Artemis program, including the first female astronaut to set foot on the Moon. Through exploring the Moon further, building on what we've already learnt, and working collaboratively with commercial space companies and international partners, the hope is to initiate a sustainable means of lunar exploration, and then expand the final frontier and send astronauts to Mars.

Space exploration has opened up the possibility of more than just detailed study of our closest celestial neighbours. We've sent spacecraft out to relatively small and mysterious space rocks in our solar system and are even bringing some of that material back. Because asteroids and comets are like time capsules, changing very little throughout history, studying them gives us clues about what the early Solar System was like when they formed. Some notable missions include:

- The Rosetta-Philae mission which was the first to orbit and land on a comet in 2014 (comet 67P/Churyumov-Gerasimenko). Its surface had ammonia, hydrogen cyanide, and hydrogen sulphide – a toxic concoction that would smell like strongly odoured urine, almonds, and rotten eggs!
- The Dawn mission which launched in 2007 investigated the two largest protoplanets in the main asteroid belt – Ceres and Vesta. Ceres, also considered a dwarf planet, displayed concentrations of salt deposits likely left behind when a previous water ocean froze, indicating that there may be places other than icy moons which may have had oceans in their past, and potentially still do.

- OSIRIS-Rex, an asteroid return mission by NASA. In October 2020 it briefly touched down on the near-Earth asteroid Bennu, and having successfully collected surface material, it is due to arrive back to Earth with the samples in 2023.
- Hayabusa, the first successful asteroid return mission developed by JAXA. Having collected samples from asteroid Itokawa, it returned in 2010. Hayabusa 2 visited near-Earth asteroid Ryugu in 2019, and returned with samples from it in December 2020. This mission also included four small rovers that could investigate the asteroid's surface providing more context about the environment and geology of the samples collected. They used a special mechanism to hop around like small solar-panelled jumping spiders, but have now run out of power.

- New Horizons in 2015 became the first spacecraft to fly-by Pluto, roughly 5 billion kilometres away from the Earth. This mission provided scientists with the most detailed images of Pluto and showed a 1,000-kilometre-wide heart-shaped glacier of frozen nitrogen – the largest in the Solar System. The spacecraft has since continued venturing further out of the Solar System. It flew by another Kuiper Belt object (KBO) called Arrokoth on New Year's Day of 2019. As it continues to leave our solar system there may be opportunity for another KBO flyby.

It's hard to define where the edge of the Solar System is – we often think about it as a limit of the Sun's influence. This could mean its light, or even its gravity, which get weaker the further away you go, but they don't ever reach zero. Another way to think about this is where the Sun's solar wind, which blows against the traces of gas and dust in space, is slowed down, forming a boundary. Within this boundary known as the heliosphere, we experience the Sun's solar wind and beyond it we can consider that we've left the Solar System.

To date, two operational spacecraft have left our solar system – Voyager 1 and Voyager 2. These spacecrafts launched from Earth in 1977 and have been travelling out ever since. Both missions, now over 18 billion kilometres from the Earth, are exploring interstellar space. But on their way out of the Solar System they gave us glimpses of our outer planetary neighbourhood and have sparked further exploration of two giant planets in our solar system in particular.

The Voyager probes gave scientists at the time new insights about the gas planets, with Voyager 1 flying by Jupiter and Saturn, and Voyager 2 going on to observe

Uranus and Neptune too. In fact, in 1989, Voyager 2 became the first spacecraft to observe our most distant planet, Neptune. Together the Voyager pair took 33,000 images of Jupiter and its five major moons and discovered active volcanism on its moon Io – it was the first time this had been seen on another object in the Solar System. They observed Titan, Saturn's largest moon, showing in detail how it was encapsulated in a thick and hazy atmosphere.

Before the Voyager probes, there were others that crossed the asteroid belt. Pioneer 10 launched in 1972 and after emerging from its journey across the asteroid belt, it became Jupiter's first spacecraft visitor. Pioneer 11 followed shortly, becoming humanity's maiden robotic encounter with Saturn.

Some of the most exciting missions have been sent to these two planets since the early flybys. The Galileo spacecraft entered

Jupiter orbit in 1995 and spent about eight years studying the Jovian system and its four Galilean moons, discovering the possibility of liquid water beneath the surfaces of a few of these satellites. In 2016 the Juno spacecraft arrived at the king of the planets with the aim of studying its composition, features, and properties in unprecedented detail. Juno is the only solar-powered deep space mission. Rather than using a nuclear fuel source, it's able to generate the power it needs thanks to its trio of 3-metre-wide by 9-metre-long solar wings – the largest arrays ever used on a planetary probe. With the yearning for an answer to the possibility of life beyond the Earth, missions are in development to study the Galilean moons of Europa, Ganymede and Callisto up-close. With liquid water below their surfaces, they could be habitable environments and ESA's JUpiter ICy moons Explorer (JUICE) mission and NASA's Europa Clipper mission will endeavour to find out. They are due to launch in the 2020s, but it will be a few years before they arrive at their destination and start sending back data that will hopefully leave us feeling incredibly excited and with more questions than before.

A mission that evokes a feeling of marvel at human endeavour, which also provided countless moments of surprise, intrigue, and wonder, is NASA's flagship mission to Saturn, Cassini-Huygens. Arriving at the jewel of our solar system in 2004, this spacecraft spent over 13 years orbiting the planet. It discovered new moons around Saturn, captured images of the hexagonal storm feature (which is twice the size of Earth) churning at its north pole, and detected icy plumes spraying out from its moon Enceladus leading to the understanding that it harbours a global water ocean under its surface. It even solved the mystery about why its moon

Iapetus appears to have a dark and bright side. Perhaps even more impressive is that the mission included a lander, Huygens, which became the first probe to land on an object in the outer Solar System after it survived its descent down to the surface of Titan. This spacecraft gave us never before seen views of the icy boulder-strewn surface of Saturn's largest moon, and we now know that Titan harbours lakes and seas of liquid methane. Much like the water cycle on Earth, the methane reservoirs evaporate into the methane-rich atmosphere of Titan, which then falls back down as liquid methane rain. With a liquid substance, a thick atmosphere and bio-signatures like carbon and hydrogen, Titan is considered to be another place that may support life.

Although there's still so much to understand about our own solar system, in the last few decades we've been pushing to discover planetary systems beyond our own. One of the most influential telescopes in the field of exoplanet research has been the 2009 Kepler space telescope. An exoplanet is a planet we find orbiting another star outside of our solar system. The existence of over 4,000 exoplanets has already been confirmed and the Kepler mission helped find a significant fraction of those. Using a technique known as the 'transit method' Kepler was able to detect the tiny decrease in a star's light as a planet periodically passed in front of it – we're talking about the light received changing by a fraction of a per cent – that's pretty sensitive science. Kepler observed 100,000 stars in a very small part of the sky (about the size of your hand held at arm's length) and although the mission has ended, another has taken the baton. TESS (Transiting Exoplanet Survey Satellite), which launched in 2018, is using the transit method to look for exoplanets too, but is searching an area of the sky

400 times larger than Kepler did. It's fascinating to think that there are worlds out there that could be like the Earth. Although we can determine the distance of an exoplanet from its star which can give us an idea about its temperature, the atmosphere can play a huge role in this. We see this in our own solar system with the planets close to Earth that perhaps could have been temperate. Instead, Venus has a greenhouse-like atmosphere, making it too hot and Mars with its extremely thin atmosphere has become too cold to be habitable.

It's the impressive space observatories and telescopes that have helped us contemplate how it all fits together. NASA's four great observatories – the Spitzer Infrared Telescope, Compton Gamma Ray Observatory, Chandra X-ray Observatory, and the Hubble Space Telescope (HST) have observed space in various wavelengths of light. By observing the Universe with these different 'eyes' we've been able to probe objects, seeing facets of their behaviour or nature, each proving to be a clue that gives us a more holistic understanding of what we are seeing. As of 2021, Hubble and Chandra remain in operation, but these old boys are soon to be demoted in the pecking order when the next generation of impressive space telescopes takes to the skies. The James Webb Space Telescope (JWST) was originally planned to launch in 2018 but was postponed several times. This telescope will revolutionise astronomy as its predecessor, Hubble, did. HST observed mainly visible light with some ultraviolet and infrared observations too, but JWST has been designed to focus on long wavelength visible light (orange-red light) up to the mid infrared range. By observing these longer wavelengths of light, JWST will be able to see high redshift objects – those which are too distant and out of reach for HST. This means it will be able to look back further into the history of the Universe and observe the formation of the first galaxies. Not only that, but JWST will have higher sensitivity too. Its primary mirror that it uses to collect light will be 6.5 metres across – that's over 2.5 times wider than Hubble's mirror. There's also the Nancy Grace Roman Telescope tipped to launch in 2025. This infrared telescope, comparable in size to Hubble, will focus on delving deeper into the mysteries of the Universe's expansion and will hope to measure the effects of dark energy – the elusive energy or force that is counteracting the pulling effects of gravity, and driving the Universe apart at an accelerated rate.

On the grandest scale, scientists have also been working on mapping our galaxy and the Universe too. Gaia is an ESA space observatory launched in 2013 with a mission to chart a 3D map of the Milky Way galaxy. In December 2020, the most detailed map of our galaxy to date using Gaia data was released, with the positions of nearly two billion stars plotted. This level of detail can help us understand other galaxies too, like those mapped by the Sloan Digital Sky Survey (SDSS). Since 2000 SSDS has used the optical telescope at Apache Point Observatory in New Mexico to create the most comprehensive map of the Universe covering 35 per cent of the sky. After 20 years it has 3D-mapped the positions of more than two million galaxies spanning back to those that were present about three billion years after the Universe began.

Space exploration has truly opened up the Universe to us. It has allowed us to develop our understanding of the cosmos and our place within it, beyond what may have seemed imaginable by even scientists of the early twentieth century. Robotic missions have dominated and will likely always be our best option for exploring deep space, but the future of space exploration in our solar system is likely to be filled with much more human exploration. Though we've seen national space corporations take the lead in our endeavour to explore space, private companies like SpaceX, Boeing, and Blue Origin are a few that could make their mark on the future of space exploration. They have found profitable ways of working in the industry by developing technologies and services that can be used by government agencies and space tourists. In 2015, SpaceX demonstrated the capability of reusable rockets when, after delivering a payload of satellites into orbit, the Falcon 9 rocket gracefully touched back down on Earth – a masterful feat! SpaceX have also developed the Dragon module – the first commercial resupply vehicle to connect to the ISS. In 2020 Doug Hurley (1966–present) and Bob Behnken (1970–present) became the first astronauts to launch from American soil in almost ten years in the first ever crewed SpaceX mission. The pair launched atop of the Falcon 9 rocket and arrived at the ISS in their Dragon module, returning to Earth after more than two months in space. In an impressive feat of engineering, the Falcon 9 returned to Earth after launching the astronauts – landing vertically on SpaceX's drone ship amusingly named 'Of Course I Still Love You' – marking the first successful landing of a reusable rocket in the ocean. Since then, NASA has launched several crews of astronauts to the ISS aboard a SpaceX Dragon Crew capsule, and in 2021 (for the first time) it saw 11 astronauts and cosmonauts packed on the space station for a few days during the crew handover. With sleeping facilities on the ISS made to cater for seven, slumber arrangements got pretty inventive!

Private companies will no doubt play a huge role in driving forward the future of space exploration. Landing on Mars, looking back to the very beginnings of the Universe, and finding life beyond the Earth, may have once seemed like science fiction, but space exploration is making them a real prospect – one that seems achievable. Perhaps as technological developments bring down the cost of space travel and exploration, one day travel into space could be as simple as catching a flight. Then, each one of us could experience seeing the one home we've always known appearing beneath us, as we travel out into space and begin our own personal journey of space exploration.

Glossary

Age of Reason – a period roughly corresponding to the eighteenth century when ideas in Europe were dominated by a pursuit for knowledge. Also known as the Age of Enlightenment, this intellectual and philosophical movement pushed aside the religious, superstitious and mystical propensities of the Middle Ages (fifth to fifteenth centuries).

Amorphous solid – a solid which has a non-crystalline order, meaning that it lacks a geometrically regular form. As a result of the constituent particles not being arranged in a highly ordered structure, the particles have some mobility and can move very slowly over time.

Astrology – the study, interpretation and connection of the positions and movements of celestial bodies with our experiences and the natural world.

Astrometry – the precise measuring of the positions, motions and magnitudes (brightness) of stars and other celestial objects.

Convex – a surface that is curved or rounded outwards, like the exterior of a circle or sphere.

Electromagnetic spectrum – the entire continuum of frequencies or wavelengths of electromagnetic radiation or light, ranging from the highest frequency and shortest wavelength gamma rays, through visible light, to the lowest frequency and longest wavelength radio waves.

Geocentric – a description of the Universe with the Earth at its centre, and the

Sun, Moon, planets and stars all orbiting around it.

Heliocentric – a description of the Universe, or in modern times the Solar System, with the Sun at its centre and the Earth and planets orbiting around it.

Kármán line – the commonly accepted boundary between Earth's atmosphere and outer space, lying at an altitude of 100 kilometres above sea level.

Payload – the capacity of what a space vehicle or aircraft can carry. Often defined as the weight it can hold, and at other times described as the contents of what it's carrying: passengers, cargo, a space probe, a satellite, animals, scientific instruments, experiments, other equipment, and even extra fuel if it's being carried optionally.

Spectrum (of light) – produced by separating a ray of light, a spectrum is a band of different wavelengths of light.

In the case of visible light, it is seen as a band of colours (like a rainbow): red, orange, yellow, green, blue, indigo and violet.

Telemetry – the collection or recording of measurements and data from a remote source, and its automatic transmission to a receiver in a different location for it to be monitored and analysed.

Useful Acronyms

ALTEA – Anomalous Long-Term Effects in Astronauts, an experiment to investigate the influence of cosmic radiation and microgravity on the central nervous system of astronauts during long spaceflights.

APP – Apollo Applications Program, a project to find alternative applications and develop science-based human spaceflight missions, from the hardware initially developed for the Apollo program.

CNSA – China National Space Administration, the national agency for China coordinating its space activities.

CSA – Canadian Space Agency, the national space agency of Canada, responsible for the Canadian astronaut program.

ESA – European Space Agency, an intergovernmental organisation established in 1975 with 22 member states including the UK.

ESO – European Southern Observatory, a European intergovernmental research organisation consisting of 16 nations, set up in the southern hemisphere.

EVA – extravehicular activity, any activity conducted by an astronaut or cosmonaut outside a spacecraft in space.

FAI – Féderátion Aéronautique Internationale or the International Aeronautical Federation, an organisation that adjudicates sporting aviation events and certifies aviation world records.

HST – Hubble Space Telescope, also known as Hubble, one of the largest and most versatile research tools for astronomers.

ISS – International Space Station, a space station launched into low Earth orbit in 1998.

JAXA – Japan Aerospace Exploration Agency, the Japanese national aerospace and space agency.

JUICE – JUpiter ICy Moons Explorer, an ESA interplanetary orbiter mission intended to explore Jupiter along with the three icy Galilean moons: Europa, Callisto and Ganymede.

JWST – James Webb Space Telescope, an orbiting infrared observatory and the successor of the Hubble Space Telescope.

KBO – Kuiper Belt Object, a space rock found in the torus of icy bodies just beyond the orbit of Neptune.

LEO – low Earth orbit, an Earth-centred orbit with an altitude between roughly 160–2000 kilometres above the Earth's surface.

LES – launch escape system, a safety system for the crew in a space capsule, which rapidly separates the capsule from the launch vehicle (rocket) in the case of an emergency.

MOL – Manned Orbiting Laboratory, a project that was part of the United States Air Force human spaceflight program in the 1960s.

NACA – National Advisory Committee for Aeronautics, a US federal agency undertaking, establishing, and promoting aeronautical research between 1915 and 1958.

NASA – National Aeronautics and Space Administration, an independent agency of the US federal government responsible for the civilian space program.

SDSS – Sloan Digital Sky Survey, one of the largest astronomical surveys that uses spectroscopy to measure the distance of numerous deep space objects.

TESS – Transiting Exoplanet Survey Satellite, a NASA space telescope launched in 2018, designed to search for exoplanets.

USSR – Union of Soviet Socialist Republics, a federal socialist state comprising of Russia and neighbouring countries that existed from 1922 to 1991. Also known as the Soviet Union.

WAC – Without Attitude Control, this second stage of the Bumper-WAC rocket had no stabilisation or guidance system, so was not able to control its attitude or orientation.

Royal Observatory
Greenwich Illuminates

Stars
by Dr Greg Brown
978-1-906367-81-7

Planets
by Dr Emily Drabek-Maunder
978-1-906367-76-3

Black Holes
by Dr Ed Bloomer
978-1-906367-85-5

The Sun
by Brendan Owens
978-1-906367-86-2